GEOMETRY
An Intuitive Approach

Meridon V. Garner
B. G. Nunley

North Texas State University

GOODYEAR PUBLISHING COMPANY, INC.
Pacific Palisades, California

Library of Congress Catalog Card Number: 70-144838

ISBN: 0-87620-350-0

Y-3500-9

Current printing (last number):

10 9 8 7 6 5 4 3 2 1

Printed in the United States of America

CONTENTS

PREFACE

The usual mathematical curriculum for elementary teachers generally begins with a course whose materials develop "the structure of the number system," followed by a course in selected topics from "informal geometry." While each of these are primarily mathematics courses, each must include applications which are related to the mathematical content and are selected from everyday situations. This is particularly true when considering the study of "informal geometry," and the particular population for which these materials are written.

In planning such a course, we found that there is no textbook available that is both mathematically correct, and fulfills the needed activities and applications for elementary teachers. Our experience over the last seven years indicate that the topics in this text will provide a background in "proper" mathematics and reinforce this knowledge by exhibiting an immediate application for the topic.

We have found that the concepts of *geometry* can be easily integrated within the mathematics curriculum for elementary teachers as follows. First, the topics in intuitive geometry can be used to complement the texts that are primarily concerned with the "structure of the number system" such as *Mathematics for Elementary Teachers*. This is particularly true when considering a six semester-hour sequence. Second, *Geometry* may be used for a three semester-hour course, as the text itself will "stand alone." Sufficient materials are provided to adequately prepare an elementary teacher for the classroom.

Following the last chapter, *Answers for Selected Problems* are provided for the readers' use. Also, the publishers can provide the instructor with an answer book containing solutions for all problems.

We wish to thank North Texas State University for providing us with research materials and other equipment necessary to prepare this book. Thanks are also due to Mrs. Cheryl Williams and Mrs. Carolyn Myers for their valuable contributions to the preparation of the manuscript.

1

REVIEW OF SETS

Almost all elementary mathematics texts begin with a study of sets and set operations, thereby claiming to be "modern." Most of these texts fail to emphasize the importance of the use of the ideas of sets within the framework of mathematics. Most of these texts also emphasize the study of sets per se, and fail to emphasize the ideas of sets when applied within the framework of the mathematics under consideration.

We assume the reader knows the meaning of the symbol "=". However, to avoid confusion, whenever this symbol is used as a connective, it should be interpreted that two names are stated and these two names are actually names for the same thing. For example, the expression "3 + 4 = 7" means "3 + 4" is a name, "7" is a name, and, in fact, they name the same thing. The symbol "=" is read "equal" or "equals." To make the statement "is not equal" we use the symbol "≠". Thus 5 + 3 ≠ 9 is a true statement.

The idea of *set* and operations on sets form the central theme when studying geometry, for geometry is a study of a particular set of points and of the various subsets of this *given* set. Hence, a quick review of the basic properties of sets should be undertaken.

First, a *set* is an idea, i.e., it is a primitive term and can only be defined by the use of other primitive terms:

Many attempts have been made to define *set*, but mathematicians no longer attempt this. We will think of a set as a "collection" of things.

From your earlier studies, you should remember the convention that "braces" are used to enclose a certain collection so that this collection will be considered as a set. If you were to consider a collection such as a notebook, a pencil, and a textbook as a set, then you could describe this set by listing these items as notebook, pencil, textbook. Remember also that each of the items listed are called *elements* or *members* of the set. These words will be used interchangeably and the meaning of "member" is precisely what you think it would be. For example, if we say "Bob is a member of the Lion's Club," we mean that Bob belongs to this particular organization.

Suppose you were asked to list the letters of the English alphabet. You should begin by writing "*A, B, C, D,*" and continue according to the pattern established. A notational device that is commonly used to mean "and continue in a like manner" is three dots, (...), placed after an element of a set. Then using this device we could list the elements of the set of letters of the English alphabet as $\{A, B, C, D,...,Z\}$. Care should be exercised in the use of this notational device so that the reader clearly understands the established pattern.

EXAMPLES:

A. $\{1, 2, 3, 5, 8,...\}$. If the "..." means "continue in a like manner", you should be able to name the next member. One pattern indicates the next member to be 13; another pattern indicates the next member to be 12.

B. $\{1, 2,...\}$ could indicate $\{1, 2, 3, 4,...\}$ or $\{1, 2, 4, 8,...\}$.

1. Listing the elements of a set:

   $\{$ Joe, Joe's dog, Joe's hat $\}$

   $\{$ Mary, Mary's doll, Sammy's turtle $\}$

   $\{a, b, c, d\}$

2. Describing the set:

   $\{$ all points in space $\}$

   $\{$ all counting numbers $\}$

   $\{$ all students in P.E. 102 $\}$

3. Using set-builder, or set-selector notation:

   $\{points \mid points$ are in space $\}$

   $\{N \mid N$ is a counting number $\}$

   $\{P \mid P$ was a pupil last year $\}$

Often, for convenience, statements are made such as "$A = \{A, E, I, O, U\}$". This sentence means that "A" is a name, "$\{A, E, I, O, U\}$" is a name, and they are names for the same thing.

If it becomes necessary to indicate that a certain thing is a member of a particular set, we could make the statement "O is a member of $\{A, E, I, O, U\}$".

The symbol "ϵ" will be used to mean "is a member of." The statement "$O \epsilon$ $\{A, E, I, O, U\}$" means that O is a member of $\{A, E, I, O, U\}$. For example,

$5\epsilon \{4, 5, 6, 7, 8\}$ means 5 is a member of the set $\{4, 5, 6, 7, 8\}$

$G\epsilon \{G, L, O, R, I, A\}$ means G is a member of the set $\{G, L, O, R, I, A\}$

Finally, there are several ideas pertaining to sets which will be used within this study of intuitive geometry.

Definition 1.1—Set Union

> For every set A and for every set B, the set of elements in set A *or* in set B is defined as the *union* of sets A and B. This is denoted $A \cup B$.

EXAMPLES:

A. $\{1, 2, 3, 4\} \cup \{1, 5, 7, 9, 11\} = \{1, 2, 3, 4, 5, 7, 9, 11\}$
B. $\{1, 2, 3, 4\} \cup \{1, 2, 7\} = \{1, 2, 3, 4, 7\}$
C. $\{1, 2\} \cup \{3, 4\} = \{1, 2, 3, 4\}$

Definition 1.2—Set Intersection

> For every set A and for every set B, the set of elements in set A *and* in set B is defined as the *intersection* of sets A and B. This is denoted $A \cap B$.

EXAMPLES:

A. $\{1, 2, 3, 4, 5, 6\} \cap \{1, 2, 3, 4\} = \{1, 2, 3, 4\}$
B. $\{1, 2, 3, 4, 5\} \cap \{1, 2, 3, 4, 5, 6, 7\} = \{1, 2, 3, 4, 5\}$
C. $\{1, 2, 3, 4, 5\} \cap \{2, 3, 8\} = \{2, 3\}$

Consider the sets $A = \{1, 2, 3, 4\}$ and $B = \{7, 8, 9, 10\}$. Note that there are no elements common to both A and B. If any pair of sets A and B have no elements in common, they are *disjoint* sets.

This illustrates that there is no element a such that $a \epsilon A$ and $a \epsilon B$. The set $A \cap B$ is the empty set, or the set that has no elements. We denote this set by $\{\ \}$ or $\emptyset$.

Definition 1.3—The Empty Set

> If A and B are disjoint sets, $A \cap B = \{\ \}$.

EXAMPLES:

A. $\{1, 2, 3, 4\} \cap \{9, 10, 11\} = \{\ \}$
B. $\{$all boys$\} \cap \{$all mothers$\} = \{\ \}$
C. $\{$all rivers$\} \cap \{$all airplanes$\} = \{\ \}$

Definition 1.4—Subset of a Given Set

For every set A and for every set B, if every element of A is also an element of B, then A is a *subset* of B. This is denoted $A \subseteq B$.

EXAMPLES:

A. $\{1, 2\} \subseteq \{1, 2, 3, 4, 5\}$
B. $\{2, 4, 6, 8, 10\} \subseteq \{$counting numbers$\}$
C. $\{a, b, c, d, e, f\} \subseteq \{a, b, c, d, e, f\}$

Definition 1.5—Proper Subset

The set A is a *proper subset* of set B if every element of A is an element of B, but *not* every element of B is an element of A. This is denoted $A \subset B$.

EXAMPLES:

A. $\{1, 2\}$ is a proper subset of $\{1, 2, 3, 4\}$
B. $\{a, b\} \subset \{a, b, c, d\}$

The reader should note that every set is a subset of itself. For example, $\{1, 2\} \subset \{1, 2\}$. Similarly, the empty set is considered a subset of every set.

Definition 1.6—One-To-One Correspondence

Two sets A and B are said to be in *one-to-one correspondence* if each element of A can be "paired" to an element of B and each element of B can be "paired" to an element of A in such a manner that distinct elements of A are paired to distinct elements of B and distinct elements of B are paired to distinct elements of A.

EXAMPLE:

Let $A = \{a, b, c, d, e, f\}$ and $B = \{1, 3, 5, 7, 9, 11\}$. Then one possible one-to-one correspondence is illustrated on page five.

$$\{a, \ b, \ c, \ d, \ e, \ f\}$$
$$\updownarrow \quad \updownarrow \quad \updownarrow \quad \updownarrow\updownarrow \quad \updownarrow$$
$$\{1, \ 3, \ 5, \ 7, \ 9, \ 11\}$$

By referring to the idea of one-to-one correspondence and the idea of proper subsets, we may think of infinite sets and finite sets.

Definition 1.7—Infinite Sets and Finite Sets

A set A is said to be an *infinite set* if it can be placed in a one-to-one correspondence with a proper subset of A.

EXAMPLES:

A. Let $A = \{1, 2, 3, 4, 5,...\}$ and $B = \{2, 4, 6, 8,...\}$. Then B is a proper subset of A. One possible one-to-one correspondence between A and B is:

$$\{1, \ 2, \ 3, \ 4, \ 5, \ 6, \ ...\}$$
$$\updownarrow \quad \updownarrow \quad \updownarrow \quad \updownarrow \quad \updownarrow \quad \updownarrow$$
$$\{2, \ 4, \ 6, \ 8, \ 10, \ 12, \ ...\}$$

Therefore, A is an infinite set.

B. Let $A = \{2, 4, 6, 8, 10,...\}$ and $B = \{4, 8, 12, 16, 20,...\}$. Then B is a proper subset of A. One possible one-to-one correspondence between A and B is:

$$\{2, \ 4, \ 6, \ 8, \ 10, \ 12, \ ...\}$$
$$\updownarrow \quad \updownarrow \quad \updownarrow \quad \updownarrow \quad \updownarrow \quad \updownarrow$$
$$\{4, \ 8, \ 12, \ 16, \ 20, \ 24, \ ...\}$$

Therefore, A is an infinite set.

Consider the set $A = \{a, e, i, o, u\}$. One proper subset of this set is $B = \{a, i, o, u\}$. After every attempt, you can see that it is impossible to establish a one-to-one correspondence between these two sets. If you were to consider any other proper subset of A, you would also find it impossible to establish a one-to-one correspondence between any of these proper subsets and the given set A. Therefore, we conclude that A is not an infinite set, which leads us to restate DEFINITION 1.7:

A set A is said to be an *Infinite Set* if it can be placed in a one-to-one correspondence with a proper subset of A. A set that is not an infinite set is a *finite set*.

EXAMPLES:

$A = \{1, 2, 3, 4\}$.

$B = \{$ all stars in the sky $\}$.

$C = \{$ all rocks in the Platte River $\}$.

Using the definition, you can verify that each of these sets is a finite set.

Definition 1.8—Set Complement

If set A is a subset of a given set B, then the *complement* of set A with respect to set B is the set of elements of B that are *not* elements of A. The complement of set A is designated **B ~ A**.

EXAMPLES:

A. If $B = \{2, 4, 6, 8, 10\}$ and $A = \{2, 4, 6\}$, then $B \sim A = \{8, 10\}$.

B. If $B = \{R, H, Y, T, H, M\}$ and $A = \{R, T, H\}$, then $B \sim A = \{H, Y, M\}$.

C. If $B = \{a, b, c, d\}$ and $A = \{a, b, c, d,\}$, then $B \sim A = \{\quad\}$.

Problem and Activity Set 1.1

A. THINKING ABOUT SETS

1. Name the elements of each set.
   a. The first six letters of the English alphabet
   b. The months of the year whose names begin with "M"
   c. The letters in your last name
   d. The numbers you use when you count the first ten children in the classroom
   e. The letters which are in the name of your hometown and are also in your first name
   f. The letters which are in the name of your hometown and are *not* in your first name

2. Here are some names of objects: a, Mary, Mississippi, steak, 1, ball, e, holly, Rio Grande, carrots, 2, 4, tricycle, i, o, Susie, Tom, Colorado, 6, 8, horses, elephants, u, Potomac, Bill, 12, 14, turnip.
   Select from these names the objects that are:
   a. The set of all girls' names
   b. The set of all names of rivers
   c. The set of all things to eat
   d. The set of all whole numbers less than 9
   e. The set of all toys

3. Make a picture of a set of objects. Then describe the set by saying: This is the set of _____.

4. Answer each of the following "true" or "false."
   a. Any collection of objects can be thought of as a set.
   b. Any collection of symbols can be thought of as a set.

c. $\{4, 5, a, b\}$ means the set whose elements are 4, 5, a, and b.

d. $\{4, 5, a, b\}$ means the set whose members are 4, 5, a, and b.

e. If $S = \{e, r, s, t\}$, then $s \in S$.

f. If $S = \{e, r, s, t\}$, then $\Delta \in S$.

g. In order for a collection of objects to be called a set, the objects have to be similar or alike in some manner.

B. ONE-TO-ONE CORRESPONDENCE

5. Which of the following illustrates a one-to-one correspondence?

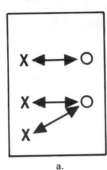 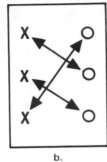 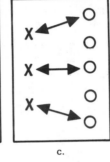 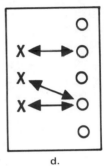

a.   b.   c.   d.

6. Illustrate a one-to-one correspondence between the set of students and their respective student numbers.

7. Illustrate a one-to-one correspondence between the set of whole numbers and the square of each whole number.

8. Let $C = \{a, b, c\}$ and $D = \{e, f, g\}$. Illustrate *every* possible one-to-one correspondence between C and D. How many are there?

9. Explain why it is impossible to illustrate a one-to-one correspondence between the set of seats in the Rose Bowl and the set of people in an overflow crowd attending the Rose Bowl.

10. Determine which of the following sets may be placed in a one-to-one correspondence with each other.

$A = \{a, b, c\}$ $E = \{1, 3, 5, 7, 9, \dots\}$

$B = \{4, 5, 6, 7\}$ $F = \{1, 2, 3\}$

$C = \{2, 4, 6, 8, 10, \dots\}$ $G = \{1, 2, 3, 4, 5, \dots\}$

$D = \{\$, \Delta, \nabla\}$ $H = \{w, t, x, y\}$

11. Using the ideas of a proper subset and one-to-one correspondence, illustrate that $\{5, 10\ 15, 20, 25, \dots\}$ is an infinite set.

C. SETS WITHIN SETS

12. The following figure shows a set within a set. The set of mathematics books is within the set of all books in the library.

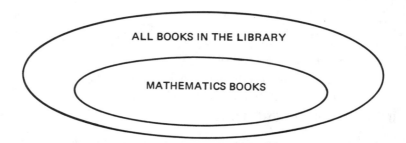

a. Make a similar drawing to show that the set of numbers 5, 10, 15, 20 is a subset of the set of all counting numbers.

b. Make a drawing to show that the numbers 1, 3, 5, 7 are a subset of the set of all odd numbers and another drawing to show that all odd numbers are members of the set of whole numbers.

13. Name one subset of each of the following sets.

a. $\{a, b, c, d, e, f, g\}$

b. $\{$ the months of the year $\}$

c. $\{$ the holidays in a year $\}$

14. Set Y is the set of all nickels. We can say that set Y is a subset of _____ .

15. Set B is the set of all even numbers. List the elements of a subset of set B.

16. Set $A = \{1, 2, 3\}$. List all subsets of set A. How many are proper subsets?

D. EQUAL SETS

17. $A = \{a, b, c\}$ $C = \{b, c, a\}$

    $B = \{c, a, b\}$

a. Does set A = set B?

b. Does set C = set A?

c. Does set B = set C?

18. Here are some sets.

$A = \{a, b, c, d\, e\}$ $D = \{g, b, d, a, e\}$

$B = \{f, a, d, b, g\}$ $E = \{b, a, c, e, d\}$

$C = \{f, a, d, b, c\}$ $F = \{e, d, g, a, b\}$

a. Set E is equal to what set(s)?

b. Set C is equal to what set(s)?

c. Are set C, set E, and set A equal?

d. Which sets are equal to set D?

19. $A = \{1, 3, 5, 7\}$. Set B is the set of all odd numbers less than 10. Which is correct?

    set A = set B or set $A \neq$ set B

20. List the elements of any set. Name this set A. Now list the elements of a set that is equal to set A.

21. Answer each of the following "true" or "false."

    a. $\{4, 5, a, b\}$ is the same set as $\{a, 5, b, 4\}$

    b. $\{4, 5, a, b\} = \{a, 5, b, 4\}$

    c. $\{1, 2, 3\} \neq \{3, 2, 1\} \neq \{2, 1, 3\}$

E. THE UNION OF SETS

22. $A = \{a, b, c, d, e, f\}$ $B = \{c, g, d, b\}$

    Which of these sets is the union of set A and set B?

    $C = \{b, d, e, f\}$

    $D = \{a, b, c, d, e, f, g\}$

    $E = \{b, c, d\}$

23. $A = \{2, 4, 6, 8, 10\}$ $B = \{3, 6, 9, 12\}$

    Write the members of the union of set A and set B.

24. Set D is the set of even numbers between 20 and 30. Set E is the set of odd numbers between 20 and 30.

    $D \cup E =$ _____

25. $F = \{\frac{1}{2}, \frac{1}{4}, \frac{1}{8}, \frac{1}{16}\}$ $G = \{\frac{1}{3}, \frac{1}{5}, \frac{1}{7}, \frac{1}{9}\}$

    Set H is the union of set F and set G. Write the members of set H.

26. Answer each of the following "true" or "false," where each of A, B, and C is a set. If you answer "false," give an example to justify your choice. (Perform operations inside parentheses first.)

    a. If $A \cup B = A$, then B is a subset of A.

    b. $A \cup B = B \cup A$

    c. $A \cup (B \cup C)$ is one set.

    d. $(A \cup B) \cup C = A \cup (B \cup C)$

F. THE INTERSECTION OF SETS

27. $A = \{1, 2, 3, 4, 5\}$ $B = \{1, 3, 5, 7, 9\}$

    Which one of these sets is the intersection of set A and set B?

    $C = \{1, 2, 3, 4, 5\}$

    $D = \{1, 3, 5\}$

    $E = \{1, 2, 3, 4, 5, 7, 9\}$

28. $F = \{a, e, i, o, u\}$ $G = \{m, a, r, y\}$

    Set H is the intersection of set F and set G. List the members of set H.

29. $A = \{3, 5, 7\}$ $C = \{13, 15, 3\}$

    $B = \{9, 11, 5\}$

Complete the following
a. $A \cap C =$ _____
b. $A \cap B =$ _____
c. $B \cap C =$ _____
d. $A \cup C =$ _____
e. $A \cup B =$ _____
f. $B \cup C =$ _____

30. Set P is the set of all odd numbers between 7 and 12. Set R is the set of all even numbers between 7 and 12.
Complete the following.
a. $P \cap R =$ _____
b. $P \cup R =$ _____

31. $A = \{2, 4, 6, 8\}$ Set B has four members.
$A \cup B = \{2, 4, 6, 8, 11, 13, 15\}$
$A \cap B = \{6\}$

Then $B =$ _____

32. Let A be the set of students with red hair. Let B be the set of students that can swim.
Then $A \cap B$ is the set _____

33. Answer each of the following "true" or "false," where each of A, B, and C is a set. If you answer "false," give an example to justify your choice.
a. $A \cap B = B \cap A$
b. If $A \cap B = A \cup B$, then $A = B$.
c. $A \cap (B \cap C)$ represents one set.
d. $A \cap (B \cup C) = A \cup (B \cap C)$
e. If $A = \{3, 4, 5\}$, $B = \{3, 4, 7, 9\}$, and $C = \{5, 9, 13\}$, then $A \cap (B \cup C) = \{3, 5\}$.
f. If $A = \{a, b, c\}$ and if $B = \{2, 3, 4, b, c, k\}$, then $A \cap B = \{b, c\}$.
g. If $A =$ the set of counting numbers less than 12 and if $B =$ the set of odd counting numbers between 6 and 16, then $A \cap B = \{7, 9, 11\}$.
h. If $A \cap B = A$, then A is a subset of B.

34. The relationship among sets A, B, and C is illustrated in the accompanying diagrams. Which of these diagrams represents $(A \cup B) \cap C$?

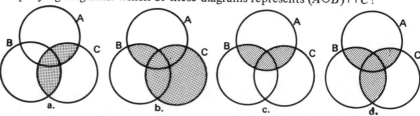

a.　　　b.　　　c.　　　d.

G. THE COMPLEMENT OF SETS

Complete the following sentences.

a. If A is the set of all students in your class and B is the set of all men students in your class, then the complement of B with respect to A is _____ .

b. If U is the set of all animals and B is the set of all horses, then the complement of B with respect to U is _____ .

c. If S is the set of all pupils and B is the set of elementary pupils, then the complement of B with respect to S is _____ .

d. If A is the set of all fish and C is the empty set, then the complement of C with respect to A is _____ .

36. If $B = \{1, 2, 3, 4, 5, 6, 7\}$ and $C = \{3, 4, 5\}$, then the complement of C with respect to B is _____ .

37. If $A = \{0, 1, 2, 3, 4, ...\}$ and $F = \{1, 2, 3, 4, 5, ...\}$, then the complement of F with respect to A is _____ .

38. In the accompanying diagrams, let X be the set of all points inside and on the rectangle and let A, B, and C denote the set of all points inside the circles as illustrated. Draw similar figures and shade the following sets.

a.

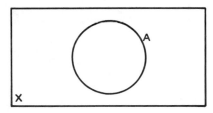

Use slanted lines to indicate the set A; use slanted lines in the other direction to indicate $X{\sim}A$.

b.

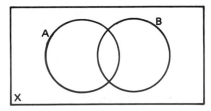

Use slanted lines to indicate the set $A{\cup}B$; use slanted lines in the other direction to indicate $X{\sim}(A{\cup}B)$.

c. Using diagram b. above, use slanted lines to indicate $A \cap B$; use slanted lines in the other direction to indicate $X \sim (A \cap B)$.

d.

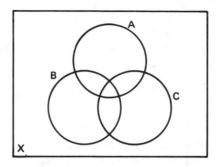

Use slanted lines to indicate the set $X{\sim}A$; use slanted lines in the other direction to indicate the set $X{\sim}B$. Draw a similar figure and shade set $X{\sim}(A{\cup}B)$. Illustrate that $(X{\sim}A)\cap(X{\sim}B) = X{\sim}(A{\cup}B)$.

39. Which set of symbols describes the shaded area of the frame?

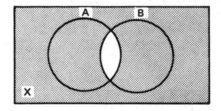

A. $(X{\sim}A)\cup(X{\sim}B)$ B. $(X{\sim}A)\cap B$

C. $(X{\sim}A)\cup B$ D. $(X{\sim}B)\cup A$

2

POINTS, LINES, SPACE, SEPARATION

2.1 THE NATURE OF INTUITIVE GEOMETRY

Many of you have heard the expression "there are more ways than one to skin a cat." We generally interpret this statement to mean that there is not just one way to do things. Such is the case in the study of geometry. One way to study geometry is to establish a deductive system in which axioms are stated and then theorems stated and proved. Another way is to create an abstract mathematical system. We will do neither of these in this study of geometry, but will rely on the ill-defined notion of "intuition." The reader's past experiences will be very valuable to him as he reads the material.

There are some truths that are "self-evident," for we can offer no justifications for their being true other than they "appear reasonable." In order for the reader to find the material meaningful, we trust he will rely upon his "intuition" and past experiences. Thus with our "intuition" in hand, we can now proceed to this study of geometry.

Often, it is inconvenient or impossible to list all the members of a set, so we resort to *describing* the set, i.e., we state in words a description that will clearly specify what things belong to the set. For example, if we wanted to consider all grains of sand on Waikiki Beach as a set, we would find it inconvenient to list the members of this set. We then resort to writing a description that clearly indicates what collection we wish to consider. We would write "{grains of sand on Waikiki Beach}" which is to be interpreted as "the set consisting of all grains of sand on Waikiki Beach at this particular instant."

Sometimes we can indicate a particular set by stating a property that all members possess. For example, if we wanted to consider the set of all left-handed students, we could list the names of these students or we could describe this set. Another way to indicate the set would be to state a property so that only things which satisfy this property are members of the set. The notation for this is to give a class name for each of the members and then to state the property which must be satisfied. Then "{$p \mid p$ is a left-handed student}" means "the set of all things, p, such that p is a left-handed student."

Hence we see that there are three ways to indicate set membership.

2.2 POINTS AND SPACE

Just as *set* is a primitive and undefined term, *point* is a primitive and undefined term. Euclid, in his *Elements of Geometry*, attempted to define a point as "that which has no depth, breadth, or width." Modern geometers have found this definition superfluous, and no longer attempt to define a point.

Intuitively, we generally think of a point as a fixed location in space, and to represent a point we sometimes use a dot (·). Of course, a dot (·) is not a point, but the smaller the dot (·), the more nearly an exact location is described.

At this juncture, we have no guarantee that points exist. If we were to study sets of points formally we would state an "existence axiom" to declare that points do exist. However, for our purposes in this book, we will assume the existence of points.

By using intuitive notions about points, we may think of space as the set of all points.

Definition 2.1—Space

Space = $\big\{$ all points $\big\}$.

Hence, *space* is an idea that each of us possesses and is based upon the fundamental concept of *point*. Our study of geometry is primarily concerned with the study of this set and certain of its subsets.

In order to refer to particular points in space, we may name the particular points under discussion. For example, points may be named by a letter and designated by a dot (·). We follow the usual convention and use capital letters for names of points. Within this text, the point named by the letter P will be distinct from the point named by the letter Q, and so on. See Figure 2.1. Thus, the point designated by the letter A (point A) will never name the same point named by the letter B (point B). Thus,

$$A \neq B,$$
$$A \neq C,$$
$$C \neq A,$$
etc.

Of interest are particular subsets of space which can be thought of as "paths" between two certain points. There are many "paths" between any two points; several are illustrated in Figure 2.2. We shall refer to any one of these paths as a curve. Therefore, a *curve* is no more than a particular set of points.

Figure 2.1

Definition 2.2—A Curve

Given two points A and B, a curve is the set of points along a "path" from A to B.

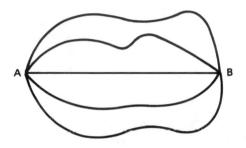

Figure 2.2

The reader should verify that not all paths are illustrated in Figure 2.2. For example, place your pencil tip on the dot that marks A and trace a path to B by lifting your pencil from the paper.

This is another "path" (curve) that cannot be illustrated here. Of all possible paths, there appears to be one that is the "most direct" path from A to B. Recall that this path is classified as a curve. Any point of this most direct path, except A and B, is said to be "between" A and B. The set containing A and B and the points which constitute the most direct path from A to B (or B to A) is called segment AB.

Definition 2.3—Segment

For every point A and for every point B, the set containing the points A and B and the points between A and B is defined as segment *AB* or *segment BA*. This is denoted $\overline{AB}$ (read segment *AB*) or $\overline{BA}$ (read segment *BA*). The points *A* and *B* are called end-points of the segment.

The reader should note that statements such as "most direct path," "between," and "to the left of," are themselves primitive expressions; each of these is an idea and involves a concept which only the reader himself can interpret. In this sense, no formal definition will be applied for these terms.

We may intuitively think of extending a segment and thereby generate another set of points. Consider segment $\overline{AB}$. If you were to go along the path from *A* to *B* and continue without termination, and then from *B* to *A* also without termination, the set of points thus generated would be called a *line*. The line *AB* is denoted $\overleftrightarrow{AB}$ and is illustrated in Figure 2.3. Very often lines are also named by lower case letters. See, for example, line ℓ in Figure 2.3.

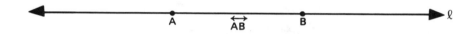

Figure 2.3

Notice in Figure 2.3 that the segment $\overline{AB}$ is a subset of line *AB*. Therefore, we may refer to the segment $\overline{AB}$ as "line segment *AB*."

We see that any two distinct points *A* and *B* determine exactly one line segment, and therefore they also determine a line. Thus, we say that *any two distinct points determine exactly one line.*

Let us investigate line ℓ in Figure 2.3 more carefully. Note that points *A* and *B* determine ℓ and hence are members of the set of points of $\overleftrightarrow{AB}$. In this sense, we think of *A* and *B* as being *collinear*.

Definition 2.4—Collinear Points

Any points which are points of the same line are called *collinear*.

In Figure 2.4, consider $\{A, B\}$, and the point $C \notin \ell$. We see immediately that $\{A, B\} \subset \ell$; hence *A* and *B* are *collinear*. But $\{A, B, C\} \not\subset \ell$, hence *A*, *B*, and *C* are *noncollinear*. But there is a line, say *m*, determined by *C* and *A*. Therefore, *C* and *A* are collinear. Likewise for *B* and *C*. See Figure 2.5.

If $\{A, C\} \subset m$ and $\{A, B\} \subset \ell$, we see immediately that $\ell \cap m = \{A\}$. Thus we conclude that if the intersection of two distinct lines is not the empty set, the intersection is exactly one point.

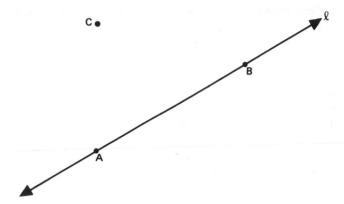

Figure 2.4

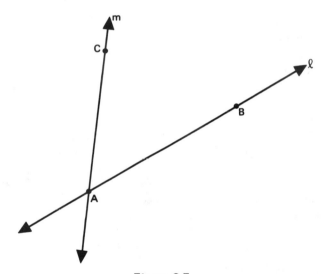

Figure 2.5

Problem and Activity Set 2.1

A. THINKING ABOUT POINTS AND SPACE

1. Which answer correctly completes the sentence? A dot (·) made with a pencil covers
   a. exactly one point
   b. several points
   c. 100 points
   d. more points than we can count
2. Which one of these most accurately describes a point?

    a. A small pencil dot (·)
    b. A very small pencil dot (·)
    c. An exact location in space
    d. A very, very small pencil dot (·)

3. Look up the word "point" in your dictionary. Look up the key words within this definition. Continue until you see a cyclic pattern. How many words are in this cycle?

4. Which answer correctly completes the sentence? In a truck load of grain, there are
    a. just as many points as grains
    b. more grains than points
    c. more points than grains

5. Which of the following represent a part of space?
    a. Your house
    b. The idea of set
    c. A building
    d. A bookcase
    e. The idea of number
    f. A piece of paper
    g. A tightly drawn string
    h. A dot (·)
    i. The air
    j. A vacuum

6. Place a book on a desk. Move the book to some other place. Does the book now represent the same set of points as before?

7. Does a log "occupy" one point of space, 1,000 points of space, or more points of space than can be counted?

8. Place a book on a desk. Label one corner of the book with a dot (·). Move the book to some other place. Did the point in space that was represented by the dot (·) "move" to the new location?

B. CURVES, LINE SEGMENTS

9. On a piece of paper, place two dots to represent two different points *A* and *B*. How many "paths" can you draw from *A* to *B*? Remember, the paths are not necessarily straight paths.

10. Trace the following figure on a separate sheet of paper.

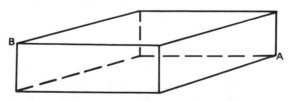

Using crayons of different colors, draw five paths from A to B. How many *different* paths do you think you could draw?

11. Mark three non-collinear points A, B, and C on a sheet of paper. Draw all possible line segments determined by these points. Label these segments. How many lines were determined? Label these lines.

12. Mark four non-collinear points A, B, C, and D. Repeat the activity as in Problem 13.

13. Mark two points A and B on a sheet of paper. Draw $\overleftrightarrow{AB}$. Draw five *other* lines containing A. Draw three *other* lines containing B.

14. Consider the accompanying figure.

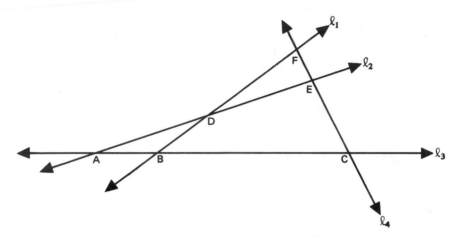

a. Name all line segments of ℓ_1, ℓ_2, ℓ_3, and ℓ_4. How many segments did you find?

b. What is $\ell_1 \cap \ell_2$? $\ell_3 \cap \ell_4$? $\ell_1 \cap \ell_3$?

15. Mark three points A, B, and C in that order on line ℓ. Color $\overline{AB}$ green. Color $\overline{BC}$ blue.

16. Repeat the activity of Problem 15, coloring only $\overline{AC}$ brown.

17. Suppose lines ℓ_1 and ℓ_2 are lines. Point A is a point of ℓ_1 and ℓ_2. Point B is also a point of ℓ_1 and ℓ_2 What can you say must be true about lines ℓ_1 and ℓ_2?

18. Here is a picture of a line.

Name all the line segments shown on the line. How many are there?

19.   Here is a drawing of a house. You cannot see part of this house.

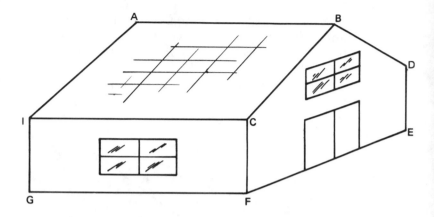

Label the corners of the foundation *E*, *F*, *G*, and *H*; some of these are labeled in the drawing. Label the unseen intersection of the roof and wall *J*. How many line segments are represented in this drawing?

20.   Here is drawing of an ordinary can.

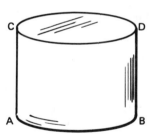

How many different curves can you find represented in this drawing?

21.   Here is a drawing of a baseball.

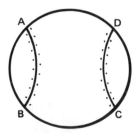

How many different curves can you find in this drawing?

22. How many different curves can you find in the accompanying figures?

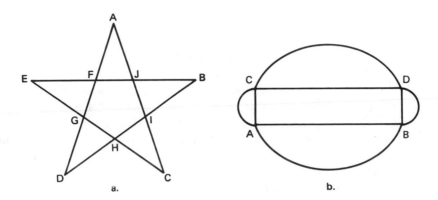

a.   b.

23. Answer each of the following "true" or "false." In each case, draw a picture or write a short sentence to justify your answer.
    a. For any two points A and B, $\overleftrightarrow{AB}$ is the same line as $\overleftrightarrow{BA}$.
    b. A segment is a subset of a line.
    c. A line is a subset of space.
    d. The set of points of a line segment is a finite set.
    e. Two distinct points determine a line.
    f. A line is a curve.
    g. Three collinear points determine exactly three line segments.

24. Using the accompanying diagram, complete the following statements.

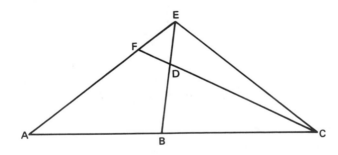

    a. ――――――, D, and B are collinear.
    b. A, ――――――, and E are collinear.

2.3 SEPARATIONS

Consider a line AB and a point P between A and B. From this we say that A and B are on "opposite sides" of P. We can think of P as a "separator" of $\overleftrightarrow{AB}$.

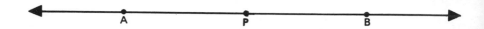

Figure 2.6

Thus, we have three disjoint subsets: (1) the point P, (2) the set of points on the A side of P, and (3) the set of points on the B side of P. Each of the sets of points in (2) and (3) is called a *half-line*. Notice that the point P belongs to neither half-line. Therefore, any point of a line separates that line into three sets as described above.

In Figure 2.7, the point P of line ℓ separates the line into two half-lines, and is referred to as the point of separation.

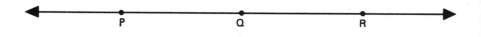

Figure 2.7

Similarly, since we think of a line as extending in opposite directions, each of the points P, Q, and R in Figure 2.8 can be thought of as a separator.

Figure 2.8

The ideas of half-lines of line ℓ and the point P of separation of line ℓ form the basis for defining a *ray*. A *ray* consists of the point P of separation together with one of the half-lines of line ℓ. In this sense, a separation point determines two rays of line ℓ.

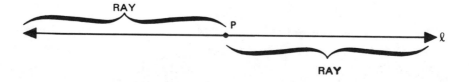

Figure 2.9

The point P is said to be the end-point of each ray as illustrated in Figure 2.9.

If each of points Q and R are points of line ℓ as shown in Figure 2.10, we may speak of ray PQ and ray PR and denote these rays as $\overrightarrow{PQ}$ and as $\overrightarrow{PR}$.

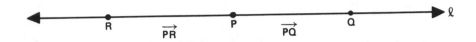

Figure 2.10

Note that point P is the end-point for *each* ray, and that $\overrightarrow{PQ} \cap \overrightarrow{PR} = P$. If $\overrightarrow{PQ} \cap \overrightarrow{PR} = P$, we say that the two rays possess a common end-point. Examples of various rays are given in Figure 2.11.

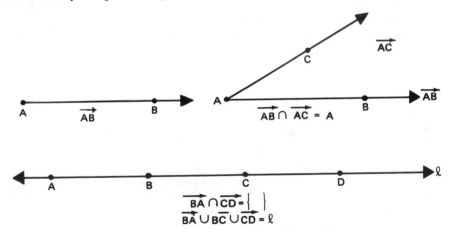

Figure 2.11

Problem and Activity Set 2.2

1. Label three points of line ℓ as A, B, and C, in that order. Color $\overrightarrow{BC}$ red. Color $\overrightarrow{BA}$ green.

2. Ray AB is represented below.

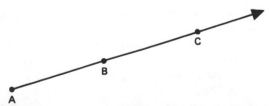

Which of the following are true?

a. $\overrightarrow{AC}$ is another name for $\overrightarrow{AB}$.

b. $\overrightarrow{BC}$ is another name for $\overrightarrow{AB}$.

c. $\overline{AB} \subset \overrightarrow{AB}$.

d. $\overline{AC} \cap \overline{AC} = \overline{AC}$

e. $\overline{AB} \cap \overline{BC} = \{\quad\}$

f. $\overrightarrow{AC} \cap \overrightarrow{BC} = \overrightarrow{BC}$

3. Mark three points A, B, and C, in that order on line ℓ. Indicate $\overrightarrow{AB}$, $\overrightarrow{BA}$, $\overrightarrow{BC}$, $\overrightarrow{AC}$, $\overrightarrow{CA}$, $\overrightarrow{CB}$.

a. Which of these are names for the same ray?

b. What is the difference between $\overline{AB}$ and $\overleftrightarrow{AB}$? Between $\overleftrightarrow{AB}$ and $\overrightarrow{AB}$?

4. Draw point A on a piece of paper. Draw four rays with end-point A. Label these rays $\overrightarrow{AB}$, $\overrightarrow{AC}$, $\overrightarrow{AD}$, and $\overrightarrow{AE}$. What is $\overrightarrow{AB} \cap \overrightarrow{AC} \cap \overrightarrow{AD} \cap \overrightarrow{AE}$?

5. Draw $\overline{AB}$ on a piece of paper. Using only the points named, name the rays containing point B. What is $\overline{AB} \cap \overrightarrow{AB}$? How many line segments are there on $\overline{AB}$ that have A as end-point?

6. Draw a line ℓ determined by points A and B. What is $\overrightarrow{BA} \cap \overrightarrow{AB}$? What is the set of points not in $\overrightarrow{AB}$?

7. Draw a line ℓ. Label four points on ℓ as A, B, C, and D, in that order. Using these points, label two rays

a. Whose intersection is one point

b. Whose union is the line

c. Whose union is not the line, but contains A, B, C, and D

d. Whose union does not contain D

e. Whose intersection is the empty set

8. Each segment and each ray in the following figure is named.

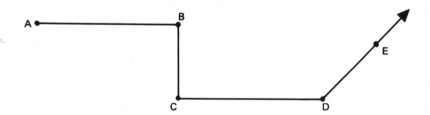

What is $\overline{AB} \cap \overrightarrow{DE}$? What is $\overline{BC} \cap \overrightarrow{DE}$? What is $\overline{CD} \cap \overrightarrow{DE}$?

9. Answer each of the following "true" or "false." Draw a picture to justify your answer.

a. For any two points A and B, $\overrightarrow{AB} = \overrightarrow{BA}$.

b. A ray is a subset of a line.

c. Two distinct points determine a ray.

d. A ray has two end-points.

e. If $\overrightarrow{AB} = \overrightarrow{AC}$, then $B = C$.

2.4 PLANES, PLANE SEPARATIONS

In the preceding section, it was determined that a point separated a line into two half-lines, from which the idea of *ray* followed.

Another idea central to the study of geometry is that of a *plane*. Intuitively, we may think of a plane as a particular set of points that lie on a flat surface which has no boundary points. Hence, plane is an abstraction of a flat surface, e.g., the set of points illustrated by the surface of a chalkboard, a sheet of paper, or a wall.

If we construct a model of a line on a flat surface which is a model of a plane, it appears that every point of the line is also a point of the plane. Since we know that any two points determine exactly one line, then it also appears that if points A and B are points of a certain plane, $\overleftrightarrow{AB}$ is a subset of that plane. In Figure 2.12, we see that if P and Q are points of plane β, then $\overleftrightarrow{PQ}$ is a subset of points of β.

Also illustrated is the case where R and S are not points of the plane α, and the line $\overleftrightarrow{RS}$ is not a subset of α. Intuitively, we also see that if two points are not in a given plane, then the line determined by these two points "pierces" the plane at one particular point.

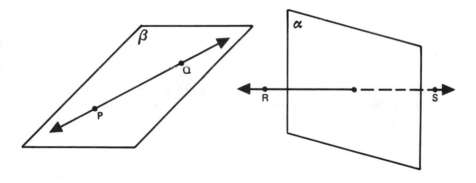

Figure 2.12

As in the case for collinear points (points that are elements of the same line), a corresponding definition may be given for points that are members of a particular plane.

Definition 2.5—Coplanar Points

A set S of points is *coplanar* if there is a plane which contains every point of S.

To illustrate the idea of coplanar points, the reader may secure three pencils and a heavy piece of paper. Place the pencils "point-up" on a tabletop and lay the heavy piece of paper on the points. Remove one of the pencils and replace it with another pencil. Again, lay the heavy paper on these pencil points. If we repeatedly perform this experiment, we see that a "plane" could contain any of the sets of three "points."

We can make the following conclusion:

"Any three non-collinear points are points of exactly one plane."

To illustrate this concept, each of *A, B,* and *C* is an element of plane α in Figure 2.13.

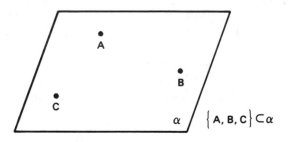

Figure 2.13

$$\{A, B, C\} \subset \alpha.$$

Similar experiments would also show that given a line ℓ and a point $C \notin \ell$, there is exactly one plane containing both of them. Furthermore, given any two lines whose intersection is not empty, there is exactly one plane containing them.

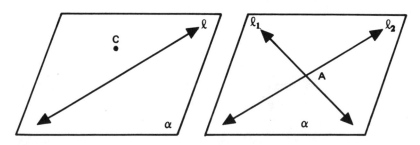

Figure 2.14

As a point separates a line into half-lines, a line separates a plane into two half-planes.

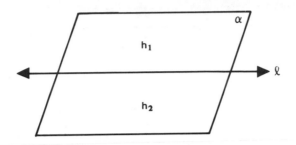

Figure 2.15

Hence, a line separates a plane into *three* disjoint subsets of the plane. Referring to Figure 2.15, two of these subsets of points are defined as half-planes, h_1 and h_2 and the line ℓ is the third set.

The idea of separation may be extended one step further. This extension involves the idea that a plane separates space into two half-spaces. Thus, the properties of separation related to points, lines, and planes are:

a. point → line → half-lines
b. line → plane → half-planes
c. plane → space → half-spaces

Problem and Activity Set 2.3

1. Does a plane as we think of it contain exactly one point or more points than can be counted?
2. Take a sheet of paper and think of it as a part of a plane. Is it possible to draw more than one line in the plane? Does a plane contain exactly one line or more lines than can be counted?
3. Think of the top of your desk as a part of a plane. Describe the location of a point which is not a point of this plane. Describe the location of a line not contained in this plane. Does a plane contain all points and all lines? Does a plane contain all points of space?
4. If the end-point of a certain ray is not an element of a certain plane, is the ray contained in the plane?
5. If the end-point of a certain ray is an element of a certain plane, is the ray contained in the plane?
6. Can four different points, no three of which are points of the same line, also be points of one plane? Must they be contained in the same plane? Illustrate your answers by drawing a picture.
7. On a level floor, why will a four-legged table sometimes "rock," while a three-legged table is always steady?
8. The accompanying figure illustrates that a line separates a plane into two half-planes, h_1 and h_2. What is $h_1 \cap h_2$?

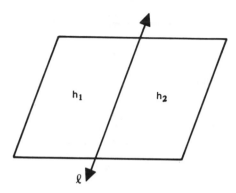

9. Consider the plane α, line ℓ of the plane, and points A, B, and C of α. Are the following the empty set, a line, or a plane?

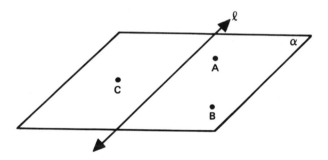

   a. $\overline{AB} \cap \ell =$
   b. $\overline{AC} \cap \ell =$
   c. $\overline{CB} \cap \ell =$

10. Consider the accompanying sketch of a house.

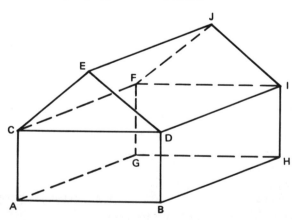

Think of the lines and planes suggested by the figure. In order to answer the following, name the lines by a pair of points and the planes by three points.

a. Name three lines that intersect in exactly one point.

b. Name a pair of planes whose intersection is exactly one line.

c. Name three planes that intersect in exactly one point.

d. Name two planes whose intersection is the empty set.

e. Name a pair of lines whose intersection is the empty set.

f. Name a line and a plane whose intersection is the empty set.

g. Name a line and a point whose intersection is the empty set.

h. Name a line and a plane whose intersection is exactly one point.

i. Name *four* planes whose intersection is exactly one point.

11. Consider the accompanying figure, which illustrates half-planes h_1 and h_2, $A \in h_1, B \in h_2$.

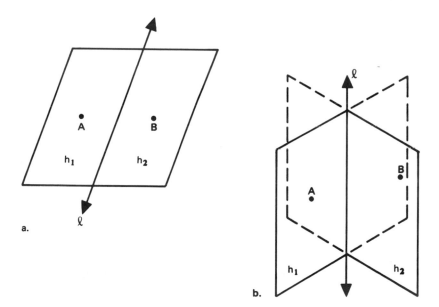

a.

b.

In Figure a, does $\overline{AB} \cap \ell = \phi$? In Figure b, does $\overline{AB} \cap \ell = \phi$? In Figure a, is $h_1 \cup h_2$ a plane? In Figure b, is $h_1 \cup h_2$ a plane?

12. Is it true that for half-planes h_1 and h_2, $h_1 \cup h_2$ is always a plane?

13. Let A and B be *any* points of half-plane α, as in the accompanying figure.

a. What is $\overleftrightarrow{AB} \cap \ell$?

b. Must $\overrightarrow{AB} \cap \ell$ be one point?

c. May $\overleftrightarrow{AB} \cap \ell = \{ \ \}$?

Draw a picture to verify your answers.

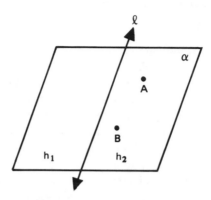

14. Let line ℓ separate plane α into half-planes h_1 and h_2. Let point A be an element of h_1 and point B be an element of h_2. We call the two half-planes into which ℓ separates α the *sides* of ℓ. We indicate the sides of ℓ by referring to the A—side of ℓ or the B—side of ℓ. The line ℓ is not in either half-plane.

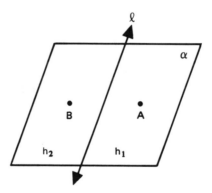

a. In the accompanying figure, is the A—Side of ℓ the same as the B—side of ℓ?

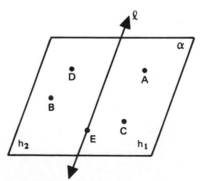

b. What is $\overline{AC} \cap \ell$?

c. What is $\overline{EC} \cap \ell$?

d. Is $\overline{AD} \cap \ell$ the empty set?

e. Is $\overline{BD} \cap \ell$ the empty set?

2.5 PARALLEL LINES AND PARALLEL PLANES

Our study has been mainly concerned with defining a particular set of points and lines which lie within the same plane. If two or more lines are subsets of the same plane they are said to be *coplanar*.

However, by investigating models of lines encountered in the everyday world, we immediately conclude that not all lines are subsets of the same plane. If two lines are not subsets of the same plane, they are said to be *skew*.

Definition 2.6—Skew Lines

If ℓ_1 and ℓ_2 are non-coplanar lines, then they are called *skew* lines.

If we consider any two coplanar lines, ℓ_1 and ℓ_2, we can immediately conclude that

A. $\ell_1 \cap \ell_2 = \{\quad\}$

or

B. $\ell_1 \cap \ell_2 \neq \{\quad\}$

If the intersection of two coplanar lines is the empty set, it appears that "they never meet," since they have no point in common. This is the notion of being *parallel*.

Definition 2.7—Parallel Lines

If ℓ_1 and ℓ_2 are coplanar lines and if $\ell_1 \cap \ell_2 = \{\quad\}$, then ℓ_1 and ℓ_2 are said to be *parallel*. To denote that ℓ_1 is parallel to ℓ_2 (and vice versa), the notation $\ell_1 \parallel \ell_2$ is used.

We may also think of planes whose intersection is the empty set. Let the chalkboard at the front of your classroom represent a portion of the points of a

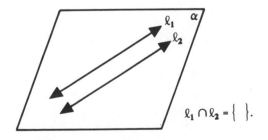

plane, then think of the wall in the back of the classroom as representing the points of another plane. It appears that these two planes will "never meet," since they have no points in common. This is the notion of parallel planes.

Definition 2.8—Parallel Planes

If α_1 and α_2 are planes and if $\alpha_1 \cap \alpha_2 = \{ \ \}$, then α_1 and α_2 are said to be *parallel*. To denote that α_1 is parallel to α_2 (and vice versa) the notation $\alpha_1 \parallel \alpha_2$ is used.

Next, let us think of two planes in space whose intersection is not empty. Does the intersection contain more than one point? Note that the planes determined by the front wall and a side wall of your room intersect in more than one point. This is illustrated in Figure 2.17.

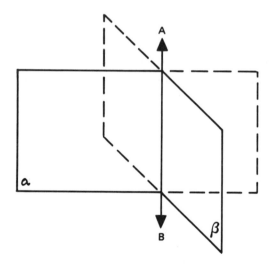

Figure 2.17

We may think of the point A as an element of α and as an element of β. Similarly, we may think of point B as a point of α and β. Now consider line $\overleftrightarrow{AB}$. We see that $\overleftrightarrow{AB}$ is a subset of α; $\overleftrightarrow{AB}$ is also a subset of β; it follows that $\overleftrightarrow{AB}$ is a subset of *both* α and β. We may conclude that in this case the intersection of α and β is *exactly one line*. We may conclude that if the intersection of two planes is not the empty set, then the intersection is a line.

Problem and Activity Set 2.4

1. Consider the point A and plane α in the accompanying figure. Point A is not an element of α. Using a ruler, construct a line $\overleftrightarrow{AB}$ such that $\overleftrightarrow{AB} \cap \alpha = \{ \ \}$. If $C \in \alpha$, construct a line $\overleftrightarrow{AC}$ so that $\overleftrightarrow{AC} \cap \alpha = C$.

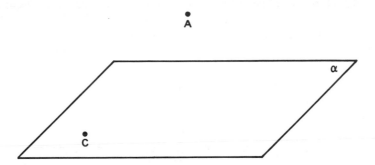

2. Describe two pairs of skew lines suggested by the edges of your classroom.
3. Fold a piece of stiff paper in half. Stand the paper on your desk so that the paper looks like a tent, as shown in the accompanying figure.

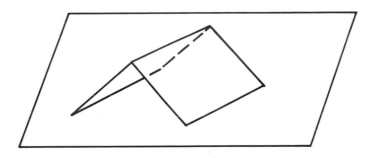

   a. Do the folded paper and the desk top suggest three planes?
   b. What is the intersection of all three planes?
   c. Consider the line suggested by the fold of the paper. What is the intersection of this line and the plane determined by the desk top?
4. Stand the folded piece of paper on the desk top as shown in the accompanying figure.

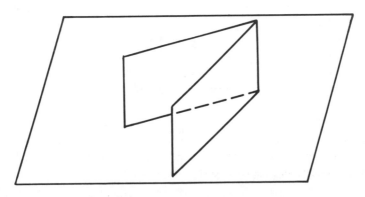

   a. Are three planes suggested?
   b. Is any point in all three planes?
   c. What is the intersection of the three planes?
   d. Consider the line represented by the fold. What is the intersection of this
      line and the plane determined by the desk top?
5. Place the folded paper on the desk top so that the fold is on the desk top, as
   shown in the accompanying figure.

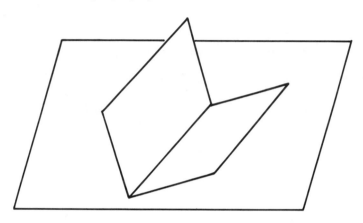

   a. Are three planes suggested?
   b. Is any point in all three planes?
   c. What is the intersection of all three planes?
   d. Consider the line determined by the folded paper. What is the intersection
      of this line and the plane determined by the desk top?
6. Consider the accompanying sketch. We may think of the "edges" as
   determining lines and the "faces" as determining planes.

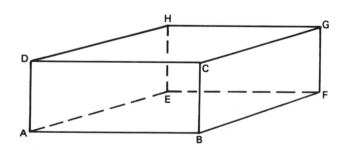

   In order to answer the following, name the lines with two points and the
   planes with three points.
   a. Name a pair of parallel lines.
   b. Name a pair of parallel planes.

c. Name two pairs of skew lines.

d. Name two planes whose intersection is exactly one line.

e. Name three planes whose intersection is exactly one point.

2.6 ANGLES

Referring to Figure 2.18, consider the points A and B and $\overleftrightarrow{AB}$. Let C be a point such that $C \notin \overleftrightarrow{AB}$. Next, consider $\overrightarrow{AC}$. Note that $\overrightarrow{AB} \cap \overrightarrow{AC} = A$, that is, A is common to both rays and A is the end-point of both rays.

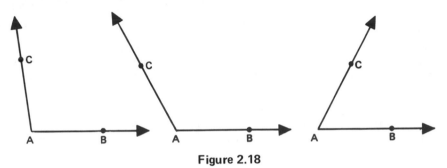

Figure 2.18

Definition 2.9—An Angle

> The union of two rays which have a common end-point and are not rays of the same line is called an *angle*.

Since $\overrightarrow{AB} \subset \overleftrightarrow{AB}$ and $C \notin \overleftrightarrow{AB}$, then $\overrightarrow{AB}$ and C belong to exactly one plane. Also, since $\overrightarrow{AC} \subset \overleftrightarrow{AC}$ and $B \notin \overleftrightarrow{AC}$, then $\overrightarrow{AC}$ and B belong to exactly one plane. We may therefore conclude that $\overrightarrow{AB}$ and $\overrightarrow{AC}$ are subsets of the *same* plane.

The common end-point of the rays of any angle is called the *vertex*, and each of the rays is called a *side* of the angle. In Figure 2.19, the angle determined by the union of $\overrightarrow{AB}$ and $\overrightarrow{AC}$ has vertex A and sides $\overrightarrow{AB}$ and $\overrightarrow{AC}$. The angle itself is denoted $\angle BAC$ or $\angle CAB$, or $\angle A$ if it is clearly understood which rays are sides of the angle.

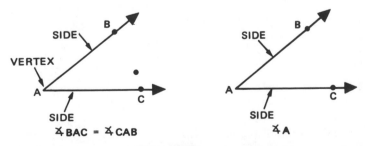

Figure 2.19

Consider ∢BAC. Also consider the half-plane on the "C-side of $\overleftrightarrow{AB}$" and the half-plane on the "B-side of $\overleftrightarrow{AC}$." The intersection of these two half-planes is called the *interior* of ∢BAC.

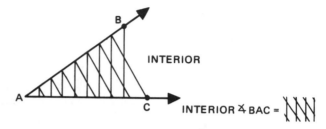

Figure 2.20

Note that $\overrightarrow{AB}$ and $\overrightarrow{AC}$ do not belong to the intersection of the two half-planes. Therefore, we see that an angle separates the plane into three disjoint subsets:
a. the set of points of the angle
b. the set of points of the interior
c. the set of all other points of the plane, called the *exterior*

Intuitively, we see that this is so because each ray of the angle extends without termination. The reader should note that the interior of the angle does not belong to the angle.

Problem and Activity Set 2.5

1. We may think of an angle as represented by two edges of your desk which meet at a corner.
   a. What represents the vertex of the angle?
   b. What represents the sides of the angle?
   c. What represents the interior of the angle?
   d. What represents the exterior of the angle?
2. Do the hands of a clock represent an angle? If so, what is the vertex of the angle? What represents the sides of the angle?
3. In each angle in the accompanying figure, name the vertex and the sides of the angle.

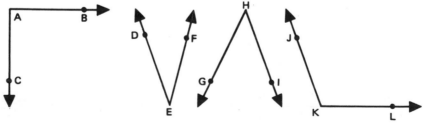

4. The accompanying figure represents ⊀*BAC*.

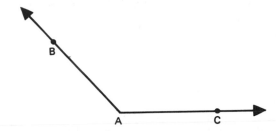

a. Choose a point on $\overrightarrow{AC}$ different from *A* and *C* and label it *D*.
b. Choose a point on $\overrightarrow{AB}$ different from *A* and *B* and label it *E*.
c. Is $\overrightarrow{AB}$ the same ray as $\overrightarrow{AD}$?
d. Is $\overrightarrow{AC}$ the same ray as $\overrightarrow{AE}$?
e. Is ⊀*BAC* the same angle as ⊀*DAE*?

5. Three points *A*, *B*, and *C* are illustrated in the accompanying figure.

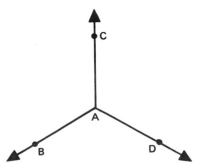

On a sheet of paper, complete the following sentences.
a. There is _____ ray containing *B* with end-point *A*.
   There is_____ ray containing *C* with end-point *A*.
c. There is _____ angle containing *B* and *C* with vertex *A*. This
   angle is labeled _____ or _____ .

6. Illustrated are three rays. Each has end-point *A*. Name three angles.

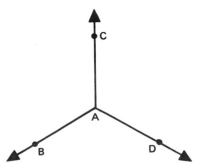

7. Mark a point *A* on your paper. Draw a picture of at least four angles which
   have the point marked *A* as the vertex. Do this by drawing five different

rays, not on the same line, with A as end-point. Choose a point different from A on each ray. Label these points T, V, W, X, and Y.

   a. Name the rays of each angle.

   b. Name each angle.

8. Label three points A, B, and C, not all points of the same line.

   a. Draw $\overleftrightarrow{AB}$, $\overleftrightarrow{AC}$, and $\overleftrightarrow{BC}$.

   b. Shade the C-side of $\overleftrightarrow{AB}$.

   c. Shade the A-side of $\overleftrightarrow{BC}$.

   d. What set is now doubly shaded?

   e. Shade the B-side of $\overleftrightarrow{AC}$.

   f. What set is now triply shaded?

9. Refer to the accompanying figure. $\angle ABC$ and points P, Q, R, and S are all in the same plane.

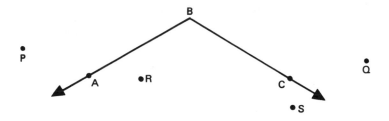

   a. Let R be a point of the interior of $\angle ABC$. Name two points of the exterior.

   b. What is $\overline{RS} \cap \angle ABC$?

   c. Is $\overline{PR} \cap \angle ABC$ exactly one point?

   d. What is $\overline{PQ} \cap \angle ABC$?

   e. Is every point of $\overline{PQ}$ in the exterior?

   f. Is every point of $\overline{RS}$ in the interior?

   g. Can you find points T and V in the exterior so that $\overline{TV} \cap \angle ABC$ is not empty?

   h. Can $\overline{SQ} \cap \angle ABC$, as illustrated above, ever be empty?

10. If possible, make sketches in which the intersection of a line and an angle is

   a. The empty set

   b. A set of two elements

   c. A set of one element

   d. A set of exactly three elements

11. If possible, make sketches in which the intersection of two angles is

   a. The empty set

   b. A set of one element

   c. A set of two elements

   d. A set of exactly three elements

12. In the accompanying figure, P lies in the interior of a certain angle. Name the angle and shade its interior.

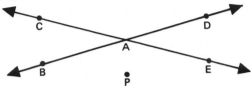

13. Is point P in the interior of $\angle ABC$? Is $\overrightarrow{BP}$ (except for B) a subset of the interior or exterior of $\angle ABC$?

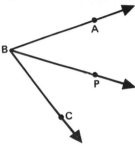

14. Answer the two questions of Problem 13 for the following case.

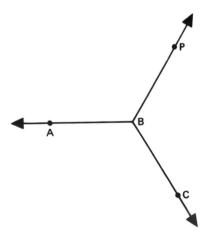

2.7 CONGRUENCE

If you were given drawings of two distinct geometric figures you might look at them and say "They look exactly alike." We could not say the two figures are equal since they are not the same set of points. However, if we could move the drawings so that one fit exactly onto the other, then we would know for sure that they were alike in every respect. Since points are fixed locations in space, we cannot move sets of points, but we can move representations of sets of points. In moving the drawings we are allowed to slide (translate), turn (rotate) or flip (transform) one of the figures.

In order to illustrate this, the reader may secure a piece of paper and a sharp pencil and then trace Figure 2.21a. After doing this, place the tracing on Figure 2.21b in such a manner that A corresponds to A', B corresponds to B', and C corresponds to C'. Similarly, trace Figure 2.21c and place the tracing on Figure 2.21d. We see that the traced figures "cover" Figures (a) and (c) and also "cover" Figures (b) and (d).

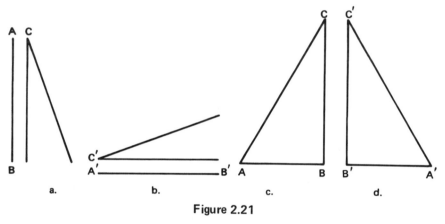

Figure 2.21

A matching scheme of this kind is called a *one-to-one correspondence* between the figures i.e., if two figures can be matched so that one figure exactly "covers" another figure, then the correspondence is called a *congruence*.

Definition 2.10—Congruence

If two figures can be matched so that one figure exactly "covers" another figure, then the correspondence is called a congruence. To denote that Figure A is congruent to Figure B, the notation $A \simeq B$ is used.

Figure 2.22 illustrates that two figures are congruent.

We see that the idea of congruence is essentially a relation; the pairs of the relation are geometric figures. What are the properties of congruence when considered as a relation among geometric figures? (1) We obviously have $a \simeq a$, since the letter "a" can name only one figure. (2) Also, if $a \simeq b$, then certainly $b \simeq a$. (3) If $a \simeq b$ and $b \simeq c$, then $a \simeq c$. These three properties describe an *equivalence* among elements of a set of geometric figures.

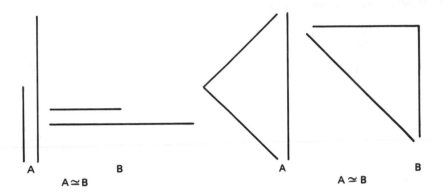

Figure 2.22

Definition 2.11—An Equivalence

If a, b, and c are elements of a set X of geometric figures, and $\simeq$ is a relation defined on X then we say $\simeq$ is an equivalence in X if

$E1$: (reflexive property): $a \simeq a$
$E2$: (symmetric property): $a \simeq b$ then $b \simeq a$
$E3$: (transitive property): $a \simeq b$ and $b \simeq c$, then $a \simeq c$

The effect of an equivalence in a set is to *partition* (sort) the elements of the set into disjoint classes. Hence, the idea of congruence may be used to partition the set of all geometric figures into disjoint classes, each figure in each class being congruent to every other figure within the class.

Problem and Activity Set 2.6

The reader should secure a piece of paper and sharp pencil and then trace some of the various figures below. After doing this, use the tracing in order to determine if one figure is congruent to other figures.

1. Which segment is congruent to $\overline{AB}$?

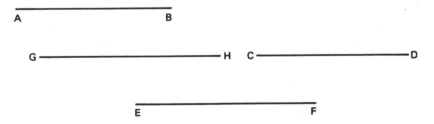

2. Which angle is congruent to ⊀*ABC*?

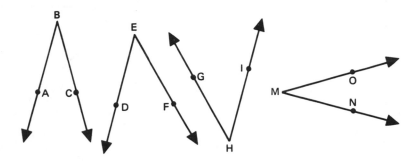

3. The accompanying figure illustrates several segments. Determine which segments are congruent.

4. Several figures are given below. Determine those pairs that are congruent.

5. If $\overline{AB} \cong \overline{CD}$ and $\overline{CD} \cong \overline{EF}$, is it true that $\overline{AB} \cong \overline{EF}$? If $\angle ABC \cong \angle DEF$ and $\angle DEF \cong \angle GHI$, is $\angle ABC \cong \angle GHI$? Explain your answer.

6. In order to determine congruence among different geometric figures, we may use tracings or, in the particular case of line segments, we may use a straightedge and a compass.

   The accompanying figure shows the steps in constructing a congruent line segment to any segment AB. Follow the steps in the figure.

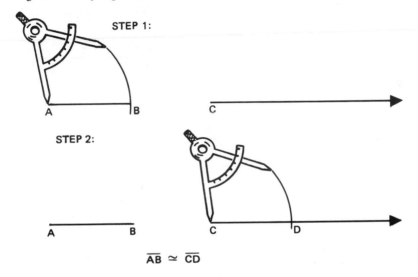

7. Similarly, we may copy any angle. In order to do this, follow the steps in the figure.

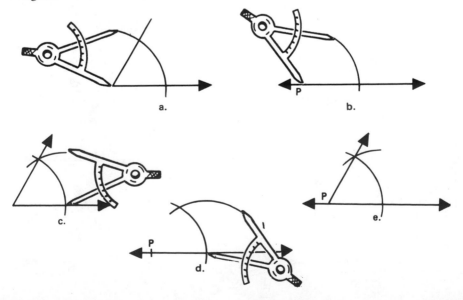

8. Given any line segment AB we can "cut" $\overline{AB}$ into two distinct congruent segments by finding a point C, so that $\overline{AC} \simeq \overline{CB}$. This is called "bisecting" the segment. In order to do this, follow the steps outlined below.

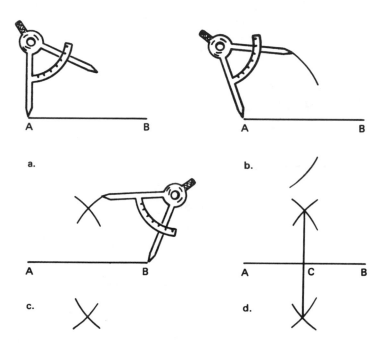

9. If we consider $\angle DAC$ there is a point P of the interior of $\angle DAC$ such that $\angle PAD \simeq \angle PAC$. The ray AP is called the bisector. Follow the steps given below to "bisect" an angle.

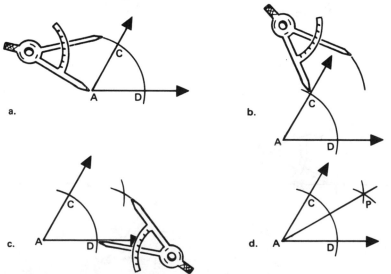

10. Using a compass and straightedge, copy each of the following figures. Then bisect each line segment and bisect each angle.

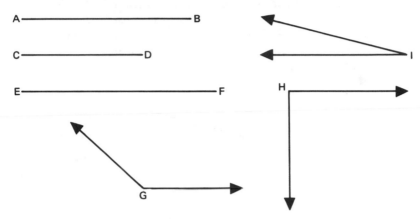

11. Copy $\overline{AB}$, below, on your paper. Bisect $\overline{AB}$ into $\overline{AC}$ and $\overline{CB}$. Then bisect $\overline{AC}$ and $\overline{CB}$, thereby constructing four congruent line segments.

A   B

12. Copy the following on your paper. Use the process of bisection to construct four congruent angles.

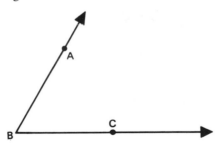

13. Can you "trisect" an angle? That is, given any ∡ABC, can you "divide" the angle into three congruent angles, using only a compass and straightedge? See if you can do this, using ∡ABC below.

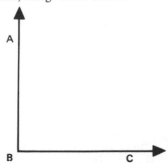

2.8 PAIRS OF ANGLES

For every two coplanar lines, ℓ_1 and ℓ_2, recall that $\ell_1 \cap \ell_2 = \{\ \ \}$ or $\ell_1 \cap \ell_2 \neq \{\ \ \}$. If $\ell_1 \cap \ell_2 = \{\ \ \}$, the lines are called parallel.

Within this section, we shall be interested in coplanar lines ℓ_1 and ℓ_2, $\ell_1 \cap \ell_2 = A$, i.e., the intersection contains exactly one point.

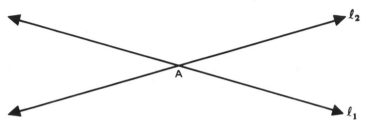

Figure 2.23

From Figure 2.23 above, we see that when $\ell_1 \cap \ell_2 = A$, four angles are determined. We may label these angles as shown in Figure 2.24.

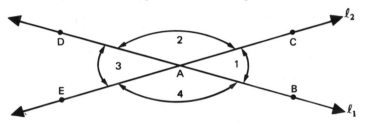

Figure 2.24

We note that rays of ℓ_1 are $\overrightarrow{AB}$ and $\overrightarrow{AD}$ and rays of ℓ_2 are $\overrightarrow{AC}$ and $\overrightarrow{AE}$. The four angles are $\angle CAB$, $\angle CAD$, $\angle DAE$, and $\angle EAB$. Also,

$$\angle BAC \cap \angle CAD = \overrightarrow{AC}$$
$$\angle CAD \cap \angle DAE = \overrightarrow{AD}$$
$$\angle DAE \cap \angle EAB = \overrightarrow{AE}$$
$$\angle EAB \cap \angle BAC = \overrightarrow{AB}$$

We see that certain pairs of angles have a common side, and all angles have the common vertex, A.

The pairs of angles which have a common side are called *adjacent* angles.

Definition 2.12—Adjacent Angles

For coplanar angles $\angle CAB$ and $\angle DAC$, if $\angle CAB \cap \angle DAC = \overrightarrow{AC}$ and if interior $\angle CAB \cap$ interior $\angle DAC = \{\ \ \}$, then $\angle CAB$ and $\angle DAC$ are called *adjacent angles*.

Hence, any two angles which have a common ray, a common vertex, and whose interiors have no common points, are called adjacent angles. Figure 2.25 illustrates adjacent angles.

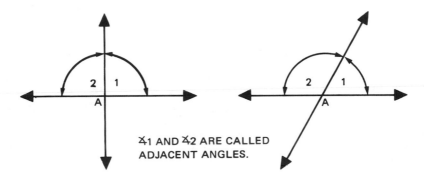

∡1 AND ∡2 ARE CALLED
ADJACENT ANGLES.

Figure 2.25

We see that two pairs of non-adjacent angles are determined by two lines intersecting in one point. These pairs are illustrated in Figure 2.26.

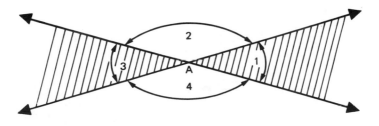

Figure 2.26

∡1 and ∡3 are non-adjacent angles, and ∡2 and ∡4 are non-adjacent angles. These two pairs of angles are called *vertical angles*.

Definition 2.13—Vertical Angles

For every coplanar line ℓ_1 and ℓ_2, $\ell_1 \cap \ell_2 = A$; the two pairs of non-adjacent angles are called *vertical* angles.

Vertical angles are also congruent angles. In Figure 2.27, ∡1 ≃ ∡3 and ∡2 ≃ ∡4. The reader should verify this by tracing ∡1 on a separate sheet of paper and then placing the traced figure on ∡3 in such a manner that the traced figure "exactly covers" ∡3. The same process may be followed with ∡2 and ∡4, in order to determine that ∡2 ≃ ∡4.

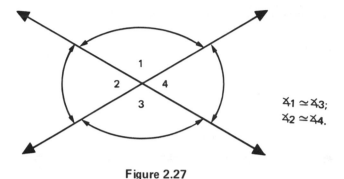

Figure 2.27

Frequently, two coplanar lines are "cut" by a third coplanar line. This third line is called a *transversal*.

Definition 2.14—A Transversal

For every coplanar line ℓ_1, ℓ_2 and t, if $t \cap \ell_1 = P$ and $t \cap \ell_2 = Q$, then t is called a *transversal* with respect to ℓ_1 and ℓ_2.

Figure 2.28 illustrates the idea of a transversal, t and the idea of different pairs of angles determined by ℓ_1, ℓ_2 and t.

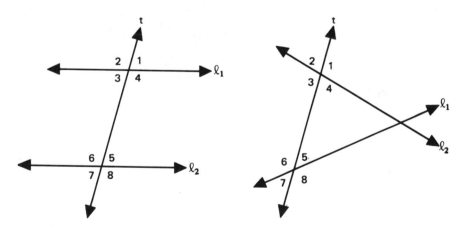

Figure 2.28

By investigation, we see that several pairs of *adjacent angles* and *vertical angles* are determined by ℓ_1, ℓ_2 and t. Several pairs of adjacent angles are $\angle 1$ and $\angle 2$, $\angle 2$ and $\angle 3$, $\angle 3$ and $\angle 4$. Two pairs of vertical angles are $\angle 1$ and $\angle 3$, $\angle 2$ and $\angle 4$. The reader may discover other pairs of adjacent angles and vertical angles by investigating Figure 2.28.

Coplanar lines ℓ_1 and ℓ_2, and transversal t of ℓ_1 and ℓ_2 also determine other pairs of angles. Additional pairs of angles determined by these three lines are called *alternate interior* angles.

Definition 2.15—Alternate Interior Angles

Let t be a transversal of ℓ_1 and ℓ_2, $t \cap \ell_1 = P$, and $t \cap \ell_2 = Q$. Let A be a point of ℓ_1 and B be a point of ℓ_2, such that A and B are on opposite sides of t. Then $\angle PQB$ and $\angle QPA$ are called *alternate interior angles*.

Note that ℓ_1, ℓ_2, and t of definition 2.13, determine two pairs of *alternate interior* angles.

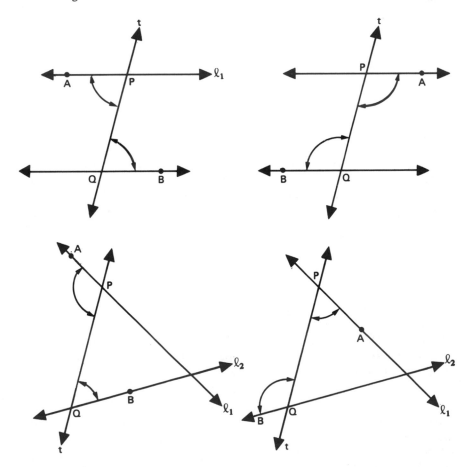

Figure 2.29. Illustration of Alternate Interior Angles

Of particular interest to the reader is the particular case when ℓ_1 and ℓ_2 are "cut" by a transversal t, and $\ell_1 \cap \ell_2 = \{\ \ \}$, i.e. ℓ_1 and ℓ_2 are parallel lines (see Figure 2.30). There are several pairs of congruent angles determined by ℓ_1, ℓ_2, and t.

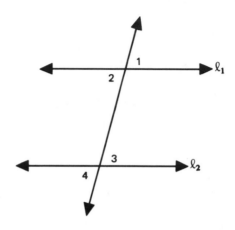

Figure 2.30

From previous discussion, we know that $\sphericalangle 1 \simeq \sphericalangle 2$ and $\sphericalangle 3 \simeq \sphericalangle 4$. But it also happens that $\sphericalangle 2 \simeq \sphericalangle 3$, or that *alternate interior angles are also congruent*. The reader should verify this in the usual manner.

We may conclude, from the transitive property of the congruence relation, that:

$$\sphericalangle 1 \simeq \sphericalangle 2 \simeq \sphericalangle 3 \simeq \sphericalangle 4$$

However, there are four angles other than those labeled in Figure 2.30. These angles are labeled $\sphericalangle 5$, $\sphericalangle 6$, $\sphericalangle 7$, and $\sphericalangle 8$ and are shown in Figure 2.31. $\sphericalangle 6$ and $\sphericalangle 7$ are alternate interior angles, and, therefore, $\sphericalangle 6 \simeq \sphericalangle 7$. The reader should verify this using a paper and pencil construction. Similarly, by using the same ideas, we may conclude that:

$$\sphericalangle 5 \simeq \sphericalangle 6 \simeq \sphericalangle 7 \simeq \sphericalangle 8$$

There are other pairs of angles determined by ℓ_1, ℓ_2, and t, provided $\ell_1 \cap \ell_2 = \{\ \ \}$, and t is a transversal. These are pairs of angles called *corresponding* angles.

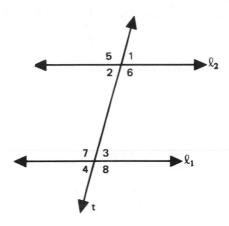

Figure 2.31

Definition 2.16—Corresponding Angles

Let ℓ_1 and ℓ_2 be coplanar lines such that $\ell_1 \cap \ell_2 = \{\ \}$ and let t be a transversal of ℓ_1 and ℓ_2 such that $t \cap \ell_1 = P$ and $t \cap \ell_2 = Q$. Let R be a point of t such that R and Q are on opposite sides of ℓ_1. Let A be a point of ℓ_1 and B be a point of ℓ_2 such that A and B are on the same side of t. Then $\angle RPA$ and $\angle PQB$ are called *corresponding angles*.

The totality of corresponding angles determined by ℓ_1, ℓ_2, and t, under the conditions established above, is shown in Figure 2.33. Note that there are four pairs of corresponding angles: $\angle A$ and $\angle A'$, $\angle B$ and $\angle B'$, $\angle C$ and C', and $\angle D$ and D'.

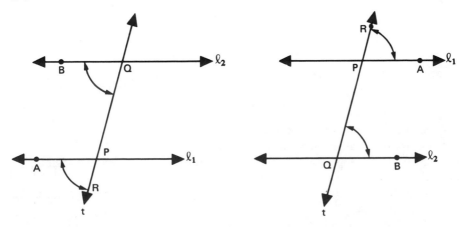

Figure 2.32. $\angle RPA$ and $\angle PQB$ are Corresponding Angles

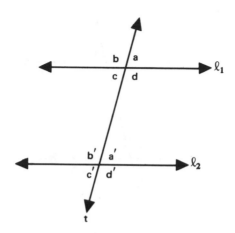

Figure 2.33

Recall that $\angle a \simeq \angle c \simeq \angle a' \simeq \angle c'$, and that $\angle b \simeq \angle d \simeq \angle b' \simeq \angle d'$. From this we can conclude that $\angle a \simeq \angle a', \angle b \simeq \angle b'$, $\angle c \simeq \angle c'$, and $\angle d \simeq \angle d'$; or we can conclude that under the condition that $\ell_1 \cap \ell_2 = \{ \ \}$, *corresponding angles are also congruent angles.*

Problem and Activity Set 2.7

1. Given the accompanying figure, answer a. – d.

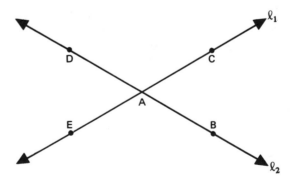

    a. Name the angles adjacent to $\angle CAB$.
    b. Name the angles adjacent to $\angle BAE$.
    c. Name the angle which with $\angle DAE$ completes a pair of vertical angles.
    d. Name another pair of vertical angles.
2. Draw a figure similar to the one given on page 53. Using this figure, answer the following.

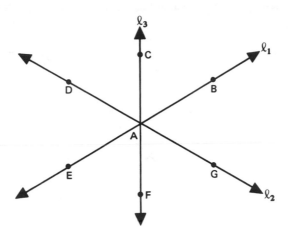

a. Name the angle which with ⊀*GAF* completes a pair of vertical angles.
b. How many pairs of vertical angles are there in the figure? Name these pairs of angles.
c. How many pairs of adjacent angles are there?
3. Draw a figure similar to the one given below. Lines $\ell_1 \cap \ell_2 = \{\ \}$. Line ℓ_3 intersects ℓ_1 at point A and ℓ_2 at point B.

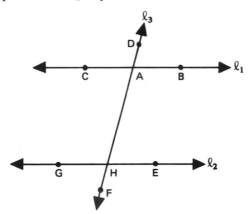

a. How many pairs of vertical angles are illustrated?
b. How many pairs of adjacent angles are illustrated?
c. Is line ℓ_3 a transversal of lines ℓ_1 and ℓ_2?
d. Is ℓ_1 a transversal of ℓ_2 and ℓ_3? Explain.
e. How many angles are determined by the three lines?
f. How many pairs of alternate interior angles are illustrated? Name these pairs of angles.

g. How many pairs of corresponding angles are illustrated? Name these pairs of angles.

h. Name every angle that is congruent to ∡DAB.

4. Draw a figure similar to the one below on your paper. Then answer the following.

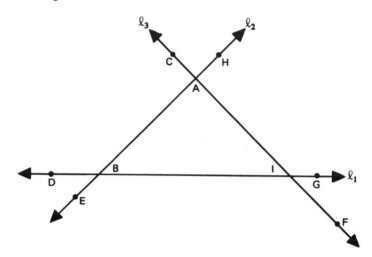

a. How many angles are determined by the three lines?
b. How many pairs of vertical angles are illustrated?
c. How many pairs of adjacent angles are illustrated?
d. How many pairs of alternate interior angles are determined by the three lines?
e. Is ℓ_1 a transversal of ℓ_2 and ℓ_3? Explain.
f. Is ℓ_2 a transversal of ℓ_1 and ℓ_3? Explain.

5. Make a figure like the one below on your paper and then answer the following.

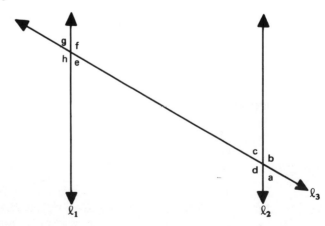

a. Name the transversal and tell what lines it intersects.
b. Name a pair of corresponding angles.
c. Name another pair of corresponding angles on the same side of the transversal as that pair selected in b.
d. Name two pairs of alternate interior angles.
e. Name all angles that are congruent to $\angle a$.
f. Name all angles that are congruent to $\angle b$.

2.9 SUPPLEMENTARY, COMPLEMENTARY ANGLES

Recall that adjacent angles are angles with a common side and a common vertex. They also contain no common interior points.

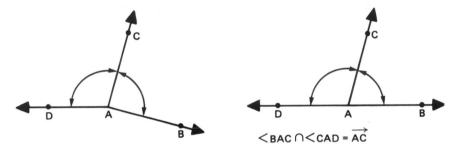

$$\angle BAC \cap \angle CAD = \overrightarrow{AC}$$

Figure 2.34

Figure 2.34a. and Figure 2.34b. each illustrate a pair of adjacent angles. But Figure 2.34b. also illustrates that:

$$\overrightarrow{AD} \cup \overrightarrow{AB} = \overleftrightarrow{DB}$$

This particular pair of angles illustrates *supplementary* angles.

Definition 2.17—Supplementary Angles

For every pair of adjacent angles, $\angle BAC$ and $\angle CAD$, if $\overrightarrow{AD} \cup \overrightarrow{AB} = \overleftrightarrow{DB}$, then $\angle BAC$ and $\angle CAD$ are called *supplementary angles*.

Figures 2.35a. and b. illustrate other pairs of supplementary angles. In some instances $\angle BAC \simeq \angle CAD$, as shown in Figure 2.35b. The reader should verify that these two angles are congruent by using the usual tracing procedure.

The angles of this particular pair of supplementary angles are *right* angles.

Definition 2.18—Right Angles

For every pair of supplementary angles, $\angle BAC$ and $\angle CAD$, if $\angle BAC \simeq \angle CAD$, then $\angle BAC$ and $\angle CAD$ are called *right angles*.

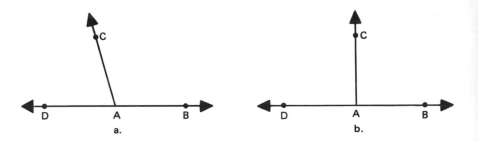

Figure 2.35

We may think of many situations in everyday life which illustrate supplementary angles and right angles. Consider the "intersection" of Avenue A and Main Street in Figure 2.36. This "intersection" determines several pairs of supplementary angles and of right angles. If we represent this map in the manner illustrated in Figure 2.37, we can list some of these pairs of angles.

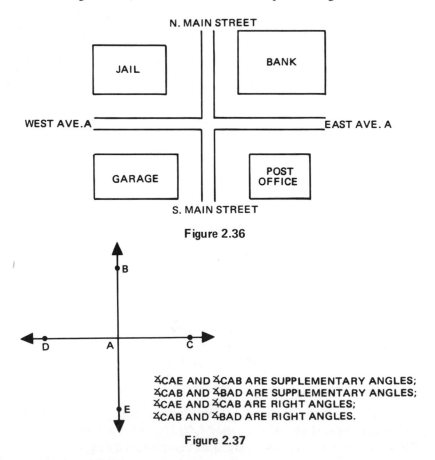

∢CAE AND ∢CAB ARE SUPPLEMENTARY ANGLES;
∢CAB AND ∢BAD ARE SUPPLEMENTARY ANGLES;
∢CAE AND ∢CAB ARE RIGHT ANGLES;
∢CAB AND ∢BAD ARE RIGHT ANGLES.

Figure 2.37

The reader should find and list other pairs of supplementary angles and right angles as above.

If two lines determine congruent adjacent angles, each angle is a right angle. These two lines are referred to as *perpendicular* lines.

Definition 2.19—Perpendicular Lines

If two lines determine congruent adjacent angles, they are called *perpendicular lines*. Perpendicular lines are indicated by the symbol ⊥.

We may also think of perpendicular rays, perpendicular line segments, and perpendiculars determined by rays and segments. Any two of the following (segments, lines, rays) are said to be perpendicular—if the lines containing these figures are perpendicular.

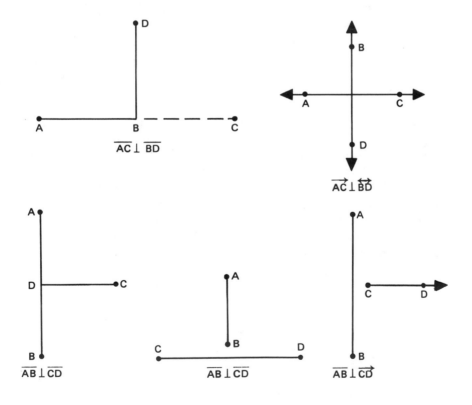

Figure 2.38

Upon occasion, two adjacent angles may "determine" a right angle. This is illustrated in Figure 2.39.

These particular adjacent angles are called *complementary angles*.

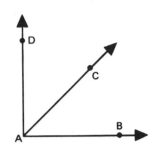

Figure 2.39

Definition 2.20—Complementary Angles

For every pair of adjacent angles, $\angle BAC$ and $\angle CAD$, if $\angle BAC \cap \angle CAD = \overrightarrow{AC}$, and $\overrightarrow{AB} \cup \overrightarrow{AD}$ is a right angle, then $\angle BAC$ and $\angle CAD$ are called *complementary angles*.

Hence, we see that both complementary angles and supplementary angles are pairs of angles; each of the definitions was based on the idea that these angles were also adjacent angles.

We may also think of supplementary angles and of complementary angles that are *not* adjacent. Consider $\angle BAC$ and $\angle FDE$ in Figure 2.40. In this figure, $\overrightarrow{AC'}$ is the *opposite* ray of $\overrightarrow{AC}$, that is $\overrightarrow{AC} \cup \overrightarrow{AC'} = \overleftrightarrow{CC'}$, and $\angle CAB$ and $\angle BAC'$ are supplementary angles. If we trace $\angle FDE$, we can place this tracing so that it "exactly covers" $\angle BAC'$. From this, we conclude that $\overrightarrow{DE}$ "covers" $\overrightarrow{AC'}$ and $\overrightarrow{DF}$ "covers" $\overrightarrow{AB}$. Since $\angle BAC'$ is supplementary to $\angle BAC$ and $\angle FDE \simeq \angle BAC'$, we conclude that $\angle FDE$ and $\angle BAC$ are also supplementary angles.

From the preceding discussion, we shall agree that for two non-adjacent angles, $\angle A$ and $\angle B$, $\angle A$ and $\angle B$ are supplementary angles if $\angle B$ is congruent to the angle that is adjacent to and is supplementary to $\angle A$. This is illustrated in Figure 2.41.

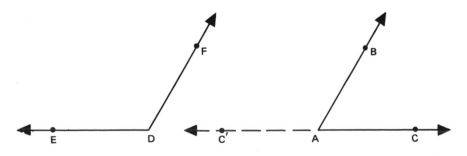

Figure 2.40

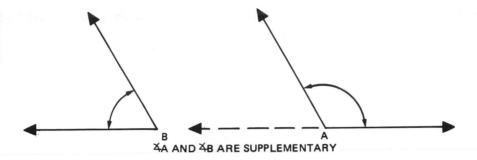

$\angle$A AND $\angle$B ARE SUPPLEMENTARY

Figure 2.41

The same arguments can be made for the relation between right angles and complementary angles. We shall also agree that for two non-adjacent angles, $\angle A$ and $\angle B$, $\angle A$ and $\angle B$ are complementary angles if $\angle B$ is congruent to the angle that is adjacent to and complementary to $\angle A$.

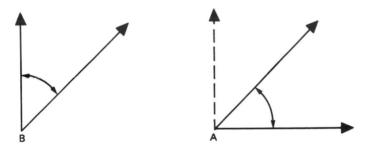

Figure 2.42

Other classifications of angles are *acute* angles and *obtuse* angles. In order to study these angles, we must consider the ideas of "greater than" and of "less than," which are also relations for pairs of angles.

Definition 2.21—"Less Than" and "Greater Than" Relations For Angles

Consider the angles ABC and $A'B'C'$. Further, consider the point D in the half-plane on the A-side of $\overleftrightarrow{BC}$ such that $\angle DBC \simeq \angle A'B'C'$. If D is in the interior of $\angle ABC$, then we say that $\angle A'B'C'$ is less than $\angle ABC$. If D is in the exterior of $\angle ABC$ then we say that $\angle A'B'C'$ is greater than $\angle ABC$. There is one other possibility for D; D could be a point of $\overrightarrow{BA}$. If this is the case, $\angle A'B'C' \simeq \angle ABC$.

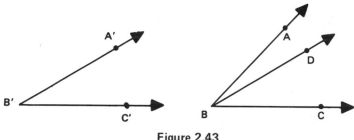

Figure 2.43

We can conclude that for any two angles, $\angle ABC$ and $\angle A'B'C'$ exactly one of the following is true:

1. $\angle ABC$ is congruent to $\angle A'B'C'$. We write $\angle ABC \simeq \angle A'B'C'$.
2. $\angle ABC$ is less than $\angle A'B'C'$. We write $\angle ABC < \angle A'B'C'$.
3. $\angle ABC$ is greater than $\angle A'B'C'$. We write $\angle ABC > \angle A'B'C'$.

We can state further that if $\angle ABC < \angle A'B'C'$ then $\angle A'B'C' > \angle ABC$.

An acute angle is an angle that is less than a right angle; an obtuse angle is one that is greater than a right angle. Each is illustrated in Figure 2.44, and the reader may verify by comparing each angle to a right angle using the usual tracing procedures.

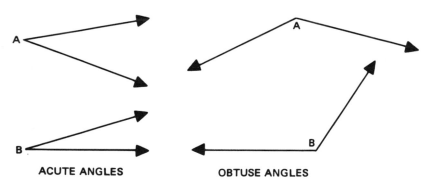

ACUTE ANGLES   **OBTUSE ANGLES**

Figure 2.44

Problem and Activity Set 2.8

1. State which of the angles appear to be right, acute, or obtuse angles.

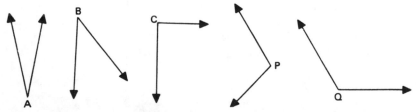

2. In the accompanying figure, name all obtuse angles which have $\overrightarrow{AH}$ as one side.

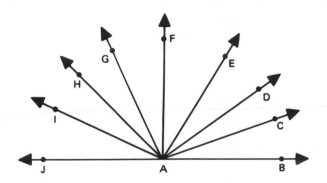

a. Name two angles which appear to be right angles.
b. Name three pairs of supplementary angles.
c. Name three acute angles which have $\overrightarrow{AE}$ as one side.
3. Which of the drawings below represent each of the following?
a. Rays which determine a right angle
b. Line segments which determine a right angle
c. Lines which determine right angles
d. A line and a ray which determine right angles
e. A line segment and a line which determine a right angle
f. A line segment and a ray which determine a right angle

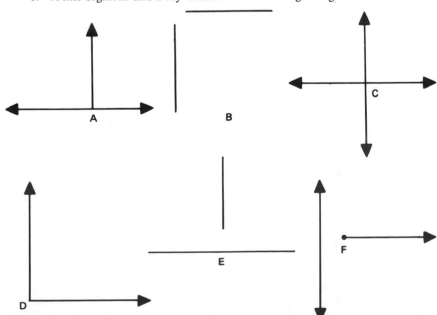

4. Find four physical representations of pairs of lines, rays, or line segments which determine right angles.

5. Find four physical representations of pairs of lines, rays, or line segments which determine angles that are *not* right angles. If they intersect, tell whether they form an acute angle or an obtuse angle.

6. Answer the following "true" or "false." You should justify your answer in each case by an illustration.

   a. An obtuse angle cannot have a complement.
   b. If two angles are adjacent, they are congruent.
   c. If two angles are supplementary, they are adjacent.
   d. If two angles are adjacent, they are complementary.
   e. Adjacent angles always occur in pairs.
   f. The supplement of an acute angle is an obtuse angle.
   g. The complement of an acute angle is an obtuse angle.
   h. The supplement of a right angle is also a right angle.
   i. If a right angle is bisected, the complementary angles determined are congruent.
   j. The supplement of an obtuse angle is an acute angle.
   k. If an angle is bisected, a pair of adjacent angles is determined.
   l. If the intersection of two lines is exactly one point, four pairs of supplementary angles are determined.
   m. If the intersection of two lines is exactly one point, four pairs of adjacent angles are determined.
   n. Two angles are complementary only if they have a common vertex.
   o. Two angles are adjacent only if they have a common vertex.
   p. Two angles are supplementary only if they have a common side.
   q. If $\angle A$ is adjacent to $\angle B$ and $\angle B$ is adjacent to $\angle C$, then $\angle A$ is adjacent to $\angle C$.

7. What kind of angle (acute, obtuse, or right) is represented by the hands of a clock when they indicate 9 o'clock? 3 o'clock? 5 o'clock? 8 o'clock? 1 o'clock?

8. We may construct a line perpendicular to a given line from any point P of the given line. Follow the steps shown in the figures to do this.

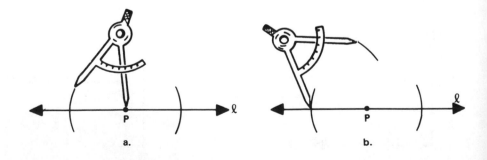

a.   b.

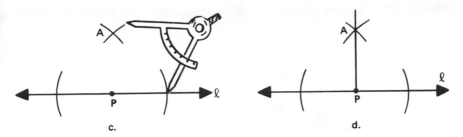

c.   d.

Draw a line ℓ and designate any point P ∈ ℓ. Follow the steps above and construct a perpendicular to ℓ at P.

9. A line segment AB is illustrated below. Copy $\overline{AB}$ and then construct a *perpendicular bisector* $\overline{CD}$ *of* $\overline{AB}$.

A ━━━━━━━━━━━━━━━━━━━━━━━━━━━━━━ B

10. Study the following construction and determine how to construct a perpendicular to a given line ℓ from a point P, P ∉ ℓ.

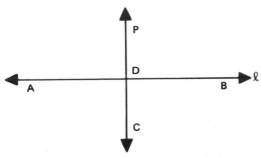

Illustrate a line ℓ and denote a point P∈ℓ. Through the point P construct a perpendicular to ℓ.

11. On a piece of paper, draw $\overline{AB}$ and illustrate bisecting $\overline{AB}$ by folding the paper so that A and B coincide. At how many points may the line segment be bisected?

12. Does it appear that a handkerchief is bisected when it is ironed?

13. How can you bisect a string? How can you cut the string into four congruent parts?

14. Cut a piece of paper so that its shape is irregular. Fold the paper to form a straightedge. Fold the paper again so that a perpendicular to the straightedge is illustrated.

15. Using an irregular piece of paper, fold the paper so that four right angles are illustrated. Fold the paper again so that the bisector of each right angle is illustrated.

16. To construct a line ℓ parallel to a given line $\overleftrightarrow{AB}$ through a point P ∉ $\overleftrightarrow{AB}$, consider the following.

a. Through the point P, draw line $\overleftrightarrow{PC}$ so that $\overleftrightarrow{PC} \cap \overleftrightarrow{AB} = C$.

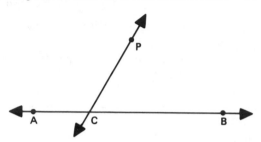

b. Using the usual methods to construct congruent angles, construct $\angle MPD \cong \angle PCB$ as shown below.

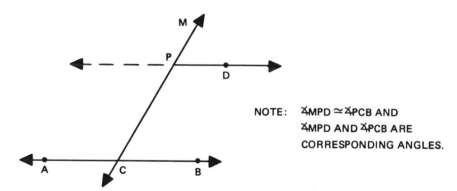

NOTE: $\angle MPD \cong \angle PCB$ AND
$\angle MPD$ AND $\angle PCB$ ARE
CORRESPONDING ANGLES.

Using the techniques above, draw any line $\overleftrightarrow{XY}$. Denote a point $M \notin \overleftrightarrow{XY}$ and construct a line through M parallel to $\overleftrightarrow{XY}$.

3

LINEAR
AND ANGULAR MEASURE

3.1 INTRODUCTION

The process of assigning numbers to "objects" or to "sets of objects" involves two distinct ideas:

a. Indexing—A part of the tabulation, or "keeping a record of" process
b. Measuring—A process which denotes characteristics of the objects

The number assigned to describe the characteristic of the object is called the *measure* of the object; the process of assigning a number to an object is called *measurement*. Careful distinction should be made between the ideas of measure and measurement. The *measure* is a number; *measurement* is the process of assigning this number. Measurement is associated with concepts such as time, velocity, rate, area, length, and volume.

Questions such as "How many students are in your class?" or "How tall are you?" have answers that are alike in one respect, for each involves a number. We may find that to answer some questions we may *count*, while to answer other questions we use *measurement*. The counting process is used to name the number of a set by establishing a one-to-one correspondence between the members of a set and a subset of the natural (or counting) numbers. Thus, we would say the number of the set $\{\triangle, \$, \odot, \boxtimes\}$ is 4:

$$
\begin{array}{cccc}
\{1, & 2, & 3, & 4, & 5\ldots\} \\
\updownarrow & \updownarrow & \updownarrow & \updownarrow \\
\{\triangle, & \$, & \odot, & \boxtimes\}
\end{array}
$$

Figure 3.1

The question "How many?" indicates you are thinking of a set of objects and wish to have a measure of this set. Such a set is called a *discrete* set, and we can determine the measure of the set by counting.

Some sets are thought of as being "all in one piece." Such sets are called *continuous* sets. For example, a wire, a road, or a water pipe may be thought of as continuous sets. Similarly, we can think of a chalkboard as a continuous set. The process of determining the measure of continuous sets is similar to that of determining measure for discrete sets. If we wish to determine the measure of a continuous set, we must first establish a *unit of measure*. The selection of this particular unit of measure depends entirely upon the characteristics of the set we wish to describe. For example, let the unit be *minute*, used in the ordinary sense. Then we would probably use this unit to describe the measure of time required to run a certain distance; we would not use this unit to describe the weight of your best friend. Thus, the unit must have the characteristic of the "same nature" as the object to which measure is assigned. Similarly, it must be possible to "move" the unit, or copy it, so that it can be used to "subdivide" the object to which measure is assigned.

The measure of any continuous set, then, depends entirely on the unit selected. To facilitate communication, it follows that some sort of "standard units" must be defined. Consider the following example.

> Tom and Jerry had identical containers and were asked to determine the measure of these containers. After experimenting, Tom replied, "The measure of my container is 6." Jerry stated, "The measure of my container is 7." It was later determined that *Tom and Jerry were both correct*.

It would appear strange indeed that identical objects could have different measures. You have probably concluded that Tom selected a unit of measure that was different from the unit selected by Jerry. To avoid such ambiguities, *standard units of measure* are determined. Particular standard units of measure will be discussed in appropriate later sections of this book.

3.2 LINEAR MEASURE

If we define the line segment $\overline{AB}$ in Figure 3.2 to be a unit of measure, we can determine the measure of other line segments. The measure of $\overline{CD}$ is assigned by tracing $\overline{AB}$ and then repeatedly placing this unit on line ℓ in such a manner that the units are "side by side," so that two consecutive segments have exactly one point in common. The segment $\overline{CD}$ is $\overline{CE} \cup \overline{EF} \cup \overline{FH} \cup \overline{HD}$. Each of these segments is congruent to $\overline{AB}$. By counting, we see that this process "subdivides" $\overline{CD}$ into four distinct units of measure. The number of units "contained within" $\overline{CD}$ is the measure of $\overline{CD}$. This is denoted "m($\overline{CD}$)=4."

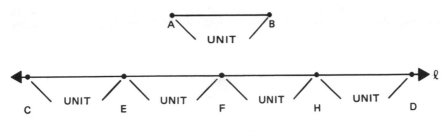

Figure 3.2

If a unit not congruent to $\overline{AB}$ had been selected in the previous example, the measure of $\overline{CD}$ could have been 2, 5, 7, 14, etc. So if a person declares that the measure of a particular line segment is 4, and the measure of another line segment is 4, we cannot conclude that the two segments are congruent. To avoid this we name the unit being used. If the measure of $\overline{AB}$ was a "glub" then the measure of $\overline{CD}$ is still 4, but we would probably say "four glubs." If we name the unit and the unit is a model of a line segment then "four glubs" is said to be the *length* of $\overline{CD}$.

Referring to Figure 3.3, if we call the unit of measure of $\overline{AB}$ "n," the unit "n" was "laid off" four times so that $\overline{CD}$ is of length "4n." In this sense, "4n" means "four segments, each congruent to the unit 'n,' were necessary to 'cover' the segment CD." Hence, we can state that:

4 is the measure of $\overline{CD}$.
n is the unit of measure.
4n is the *length* of $\overline{CD}$.

Measures of this type are called *linear measures*. "Linear" means "having the nature of a line." All measures of line segments are called linear measures.

One *standard unit* for linear measure is the *meter*. The meter was originally defined as one ten-millionth of the distance between the equator and the North Pole. Other familiar standard units of measure are the yard, foot, rod, inch, and mile.

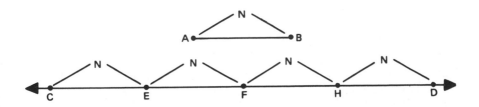

Figure 3.3

When the measure of a line segment is determined using the "inch" as the unit of measure, units one inch in length are placed "side by side" on the line segment so that any two consecutive segments have exactly one point in common.

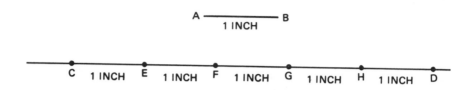

Figure 3.4

Referring to Figure 3.4, the segment CD is the union of $\overline{CE}$, $\overline{EF}$, $\overline{FG}$, $\overline{GH}$, and $\overline{HD}$, each segment being congruent to the unit of measure. If $\overline{CD}$ is exactly five inches long, this merely states that it is the union of five unit segments, each segment one inch long:

$m(\overline{CD}) = 5$
"inch" is the unit
"5 inches" is the length

However, not all measures are "exact" in the same sense noted above. Suppose we wish to determine the measure of $\overline{AB}$ in Figure 3.5, using the standard unit "inch." The reader will notice that three units of measure will not "cover" AB and that when using four units of measure, $\overline{AB}$ is more than "covered." The nature of measurement requires that a number be assigned. Therefore, the measure is given "to the nearest unit." In the example given, we would conclude that $m(\overline{AB})=4$. The reader may recognize that there is an error, but the only way to reduce the error would be use another unit of measure.

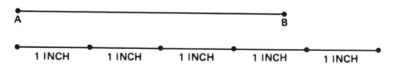

Figure 3.5

By using the usual tracing techniques, the reader can determine that the two segments on page 69 are congruent segments, i.e., $\overline{AB} \simeq \overline{CD}$. Upon further investigation, the reader may also determine that the measures of the two segments are equal. Hence we can state that "if two segments are congruent, their measures

are equal, provided the same unit is used to determine their measure."
Conversely, "two line segments that have the same measure, derived from the
same unit, are called congruent segments."

When standard units such as the "foot," "yard," "meter," or "mile" are used,
the same process is applied. In each process, the measure is determined by
placing the unit "side by side" so that two consecutive segments have exactly
one point in common. The length is then stated in terms of the particular unit of
measure.

Problem and Activity Set 3.1

1. Let the unit of measure "glub" be defined as indicated below. Determine
   the length of the following segments to the nearest "glub."

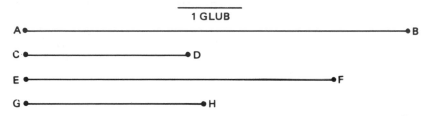

2. Determine the length of the segments above, if the unit is
   a. the inch
   b. One-fourth of one inch
   c. One centimeter (trace the segments and use the sketch in problem 5).
   d. one-tenth of one centimeter
3. If the inch is the unit, which of the segments have equal measures?
4. Which of the segments have the same length?
5. Below is a ruler scaled in *centimeters*, each centimeter divided into tenths.
   One-tenth of a centimeter is called a *millimeter*. By comparing this ruler to
   the one scaled in inches, half-inches, etc., express each length given in
   inches to the nearest centimeter; to the nearest millimeter.

     a. 1 inch   b. 2 inches

     c. 3 inches   d. 1/2 inch

     e. 1-1/2 inches   f. 3-5/8 inches

6. Using the figure of Problem 5, express each length given in centimeters to the nearest 1/8 inch.

     a. 10 centimeters   b. 5 centimeters

     c. 12 centimeters   d. 45 centimeters

     e. 92 centimeters   f. 25 centimeters

7. A pupil found a part of a broken ruler. The pupil wished to measure the line segment AB below. Explain how the pupil could determine the length of the segment.

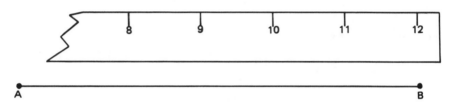

8. If $m(\overline{AB}) = 8$, construct a perpendicular bisector $\overline{CD}$, $m(\overline{CD}) = 6$, of $\overline{AB}$. Let the unit of measure be the inch. Draw $\overline{AC}$, $\overline{CB}$, $\overline{BD}$, and $\overline{DA}$. Determine, to the nearest inch, the length of these segments.

9.

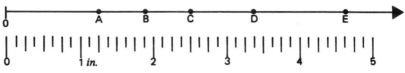

How long is

     a. $\overline{AB}$?   b. $\overline{BE}$?

     c. $\overline{BD}$?   d. $\overline{AE}$?

10. Answer the following either "true" or "false." Mark a statement "true" if it is true in every instance. Otherwise, it is "false."

     a. A unit of measure is a length.

     b. The measure of a line segment is a number.

     c. $\overline{AB} = 10$

     d. If $m(\overline{CD}) = m(\overline{AB})$, then $\overline{CD} \simeq \overline{AB}$.

     e. If the measures of two line segments are derived using different units, the segments are not congruent.

     f. If $m(\overline{AB}) \neq m(\overline{CD})$, then $m(\overline{AB}) < m(\overline{CD})$ or $m(\overline{AB}) > m(\overline{CD})$.

     g. If line segment $\overline{AB} = \overline{AD} \cup \overline{DB}$ and if $m(\overline{AD}) = m(\overline{DB})$, then D bisects $\overline{AB}$.

3.3 MEASUREMENT OF ANGLES

As noted previously, to determine the measure of an "object" the unit of measure must be of the same nature as the object to which measure is assigned. In the preceding section, we assigned numbers to line segments by using line segments as units of measure. Similarly, we determine the measure of angles using an angle as the unit of measure.

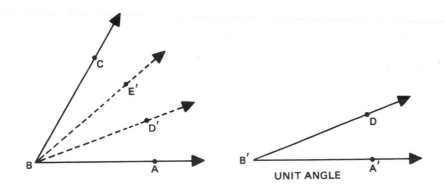

Figure 3.6

The measure of ∡ *ABC* in Figure 3.6 may be found by tracing the unit angle and then placing one side of the unit angle on $\overrightarrow{AB}$, so that *B* and *B'* coincide and the other side of the unit angle "covers" points of the interior of ∡*ABC*. Repeat the process so that *B'A'* "covers" $\overrightarrow{BD'}$ and *D* coincides with *E'*. In this manner, we see that ∡*ABC* is subdivided into three "consecutive" angles which are congruent to the unit angle and which have exactly one side in common. The measure of ∡*ABC*, using this particular unit, is 3:

$$m(∡ABC) = 3$$

Just as there are standard units of linear measure, there are standard units for assigning the measure of an angle. In order to "construct" one of these standard units, start with line ℓ and point *A* on line ℓ. Through point *A*, draw 179 rays in one of the half-planes determined by line ℓ, so that they, together with the rays of ℓ, determine 180 congruent angles. The union of these angles, together with 'their interiors, is the half-plane. Number the rays consecutively from 0 to 180, so that a number corresponds to each ray, and conversely. Any one of these congruent angles is defined as a *standard unit*; the measure of each angle is 1; the unit is called a *degree*. In Figure 3.7, only the ray corresponding to 0 and every tenth ray thereafter is illustrated.

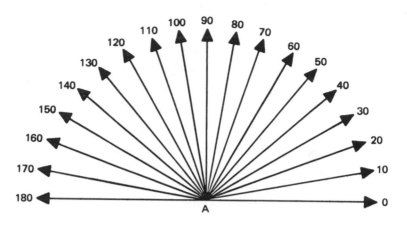

Figure 3.7

In order to assign measure to an angle using this model, place the angle on the model with one side of the angle on the ray labeled 0 and the vertex of the angle at the intersection of the rays. Then the number which corresponds to that ray along which the remaining side of the angle falls is the measure of the angle. In Figure 3.8, the measure of $\angle BAC$ is 50. The measure of $\angle BAC$ is denoted "m($\angle BAC$) = 50," and the magnitude of $\angle BAC$ is denoted 50 degrees. The symbol for degree is °, and we may say that the magnitude of $\angle BAC$ is 50°, which means m($\angle BAC$) = 50 with the unit of measure the degree.

Since it is inconvenient to "place" an angle on a model such as that in Figure 3.8, a *protractor* is used to assign measure to angles. This model is then placed on the angle instead of placing the angle on the model. To measure $\angle BAC$ in Figure 3.9,

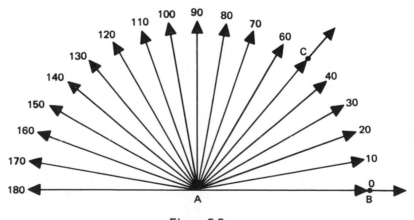

Figure 3.8

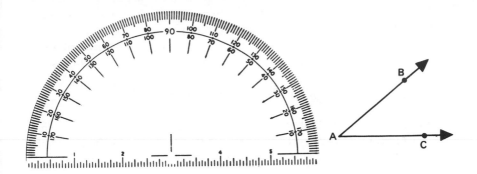

Figure 3.9

using a protractor, place the protractor on the angle so that point "▲" is on the vertex of the angle and the ray corresponding to zero lies on one side of the angle. Then the number which corresponds to the protractor ray which lies on the other side of the angle is the measure of the angle. In Figure 3.9, m(∡BAC) = 40.

3.4 MEASURES OF PAIRS OF ANGLES

When we consider the measure of pairs of angles, we may consider congruent angles, complementary angles, supplementary angles, and acute and obtuse angles.

We frequently think of congruent angles as angles that are of the same "magnitude." If we determine the measures of several pairs of congruent angles, in degrees, we find that the measures of the congruent angles are equal. We are able to say that if two angles are congruent, their measures are equal; conversely, if the measures of two angles are equal, then the angles are congruent.

Certain other pairs of angles are supplementary angles. The definition for supplementary angles has been previously stated. The reader should also notice that the sum of the measures (in degrees) of two supplementary angles is 180. Recall that the measure is a number, so we are justified in speaking of the sum of the measures. *We do not add angles, we add measures.*

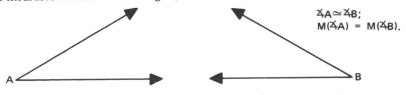

∡A ≅ ∡B;
M(∡A) = M(∡B).

Figure 3.10

Recall that supplementary angles could be adjacent angles. Also, for two non-adjacent angles, A and B, they are supplementary angles if B is congruent to the angle that is adjacent to and supplementary to A. This is illustrated in Figure 3.11. Using the usual techniques for determining the measure of angles, the reader may verify that the sum of the measures of $\angle CAB$ and $\angle DAB$ is 180:

$$m(\angle CAB) + m(\angle DAB) = 180$$

In Figure 3.11, which illustrates two supplementary, non-adjacent angles, note that $m(\angle EBG) + m(\angle EBF) = 180$. Since $\angle DAC \simeq \angle EBG$ it follows that $m(\angle DAC) = m(\angle EBG)$. By substitution, we may conclude that:

$$m(\angle DAC) + m(\angle EBF) = 180$$

Hence, we see that if two angles are supplementary, the sum of their measures is 180; conversely, we may also say that if the sum of the measures is 180, the angles are supplementary.

If two angles are supplementary and congruent, as illustrated in Figure 3.12, the two angles are right angles.

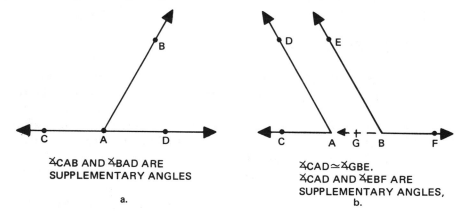

∡CAB AND ∡BAD ARE
SUPPLEMENTARY ANGLES

a.

∡CAD≃∡GBE.
∡CAD AND ∡EBF ARE
SUPPLEMENTARY ANGLES,
b.

Figure 3.11

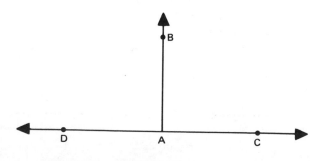

Figure 3.12

The sum of the measures of ∡*CAB* and ∡*BAD* is 180, since they are supplementary angles. Since ∡*CAB* ≃ ∡*BAD*, we may then conclude that m(∡*CAB*) = m(∡*BAD*), or that:

$$m(∡CAB) = m(∡BAD) = 90$$

Recall that complementary angles were defined in terms of right angles. Similar arguments can be made for the relation between right angles and complementary angles as were made for the relation for supplementary angles.

In Figure 3.13a let ∡*EAC* be a right angle. It follows that the sum of the measure of the complementary angles, ∡*CAD* and ∡*DAE*, is 90, since m(∡*CAE*) = 90.

In Figure 3.13b, let ∡*CAD* and ∡*DAE* be complementary angles. ∡*FBG* ≃ ∡*DAE*. Since ∡*CAD* and ∡*DAE* are complementary, m(∡*CAD*) + m(∡*DAE*) = 90. Also, since ∡*FBG* ≃ ∡*DAE*, it follows that m(∡*FBG*) = m(∡*DAE*). By substitution, we are able to determine:

$$m(∡FBG) + m(∡CAD) = 90$$

Thus, we are able to conclude that if two angles are complementary, the sum of their measures is 90. Also, if the sum of the measures of two angles is 90, the angles are complementary.

An *acute* angle is an angle that is less than a right angle, and an *obtuse* angle is an angle that is greater than a right angle. In Figure 3.14, ∡*CAE* is a right angle. Since ∡*CAB* of Figure 3.14a is an acute angle and is therefore less than a right angle, it follows that $0 < m(∡CAB) < 90$. Also, since ∡*CAH* of Figure 3.14b is an obtuse angle and is therefore greater than a right angle, we may say that $90 < m(∡CAH) < 180$.

To summarize:

    a. If *A* is an acute angle, then $0 < m(∡A) < 90$. Conversely, if $0 < m(∡A) < 90$, then *A* is an acute angle.

    b. If *B* is an obtuse angle, then $90 < m(∡B) < 180$. Also, if $90 < m(∡B) < 180$, then *B* is an obtuse angle.

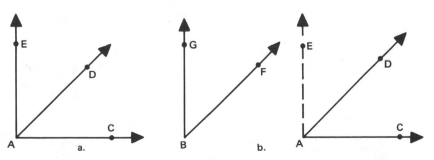

Figure 3.13

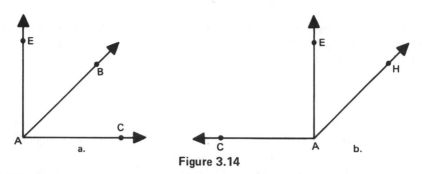

Figure 3.14

Problem and Activity Set 3.2

1. Let the angle illustrated in Figure a. be the unit "glub." Trace this figure and then determine the measure of angle A in terms of this unit.

2. Determine the measure of angle A of Problem 1, using each of the units denoted below.

3. In the following, a protractor is shown placed on a figure of several rays with end-point H.

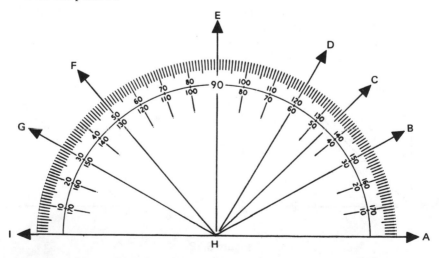

Determine the measure, in degrees, of each of the following angles

a. ∢*AHD*   b. ∢*AHF*

c. ∢*AHE*   d. ∢*BHE*

e. ∢*CHF*   f. ∢*FHG*

g. ∢*DHI*   h. ∢*DHF*

i. ∢*EHI*

j. Name three pairs of supplementary angles.

k. Name three pairs of complementary angles.

l. Name two acute angles.

m. Name two obtuse angles.

n. Name two congruent angles.

4. Given line *AB* and point *P*∊ $\overleftrightarrow{AB}$, *P* between *A* and *B*, construct a perpendicular $\overline{PC}$ to $\overleftrightarrow{AB}$, using compass and straightedge techniques. What is the measure, in degrees, of ∢*APC* and ∢*BPC*?

5. Bisect ∢*APC* and ∢*BPC* of Problem 4, again using compass and straightedge. What is the measure, in degrees, of each acute angle? Name three pairs of complementary angles.

6. In the accompanying figure, the measure, in degrees, of ∢*BAC* is 30.

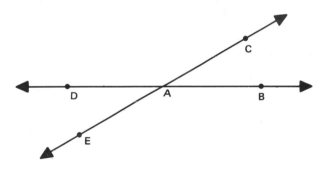

a. Name all angles congruent to ∢*BAC*.

b. Name two angles that are adjacent to ∢*BAC*.

c. What is the measure, in degrees, of these adjacent angles?

d. What is the measure of the angle which with ∢*BAC* completes a pair of vertical angles?

e. Name another pair of vertical angles shown in the figure. What is the measure of each angle of the pair?

7. The following figure is similar to that of Problem 6. The angles *A, B*, and *C* are indicated by letters in the interior of the respective angles.

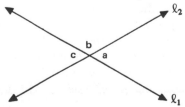

Complete the following.

a. m($\angle A$) + m ($\angle B$) =

b. m($\angle C$) + m($\angle B$) =

c. If m($\angle A$) is known, how can you find m($\angle B$)? How can you find m($\angle C$)?

8. In the following, m($\angle ABC$) = 60, where the unit is degrees. Determine the measure of every angle denoted *without* use of a protractor. $\ell_1 \parallel \ell_2$.

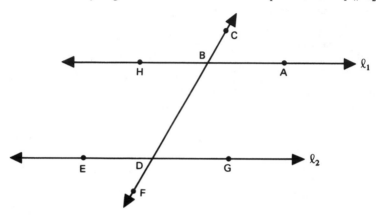

9. Copy on your paper the accompanying figure. $\ell_1 \parallel \overleftrightarrow{AB}$.

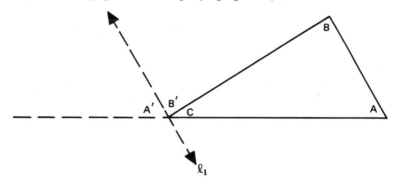

a. Is $\angle A \simeq \angle A'$? Does m($\angle A$) = m($\angle A'$)? Why?

b. Is $\angle B \simeq \angle B'$? Does m($\angle B$) = m($\angle B'$)? Why?

c. What does m($\angle C$) + m($\angle A'$) + m($\angle B'$) appear to be?

d. From the information above, does it appear that m($\angle A$) + m($\angle B$) + m($\angle C$) = 180, if the unit is degrees?

10. Copy on your paper the accompanying figure. Cut out the paper that represents the interior of the triangle. Tear off two of the corners representing vertex A and vertex B and mount the whole figure on a sheet of paper. Place vertex A and B around vertex C as illustrated. Determine the measure of $\angle A$, $\angle B$, and $\angle C$. What is m($\angle A$) + m($\angle B$) + m($\angle C$)?

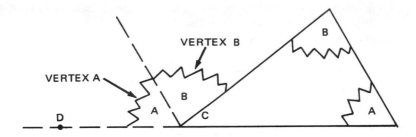

11. Draw a triangle on your paper congruent to the one represented below.

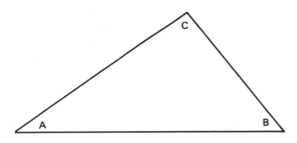

Find the mid-point of segment $\overline{AC}$, and of $\overline{BC}$, using the usual compass and straightedge techniques. Label these points D and E, respectively. Draw $\overline{DE}$.

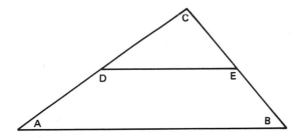

Cut out the paper that represents the interior of the triangle represented on your paper. Fold downward the portion of the triangle containing vertex C along $\overline{DE}$ so that the vertex C falls on $\overline{AB}$. Label the point where C falls on $\overline{AB}$ as F.

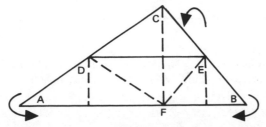

Fold the portion containing A to the right so that A falls on F. Fold the portion containing B to the left so that B falls on F. Your completed figure should represent a rectangle.

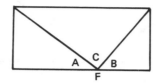

a. What appears to be true concerning the sum of the measures of $\angle A$, $\angle B$, and $\angle C$?

b. Does this experiment work with other triangles?

c. Do the results of this experiment agree with the results of Problems 9 and 10? Can you conclude, from these three problems, that the sum of the measures of the angles of a triangle is 180?

12. What is the measure, in degrees, of each angle of an equilateral triangle?

13. What is the measure of the third angle of the triangles if two of the angles of the triangles have the following measures, where the unit is degrees?
   a. 50 and 70
   b. 120 and 30
   c. 40 and 35
   d. 60 and 90

14. Suppose one angle of an isosceles triangle has a measure of 70, where the unit is degrees. Determine the measures of the other two angles. Are two answers possible?

15. Given the following line segment and angles, using the usual construction techniques, construct $\triangle ABC$. Determine m($\angle C$).

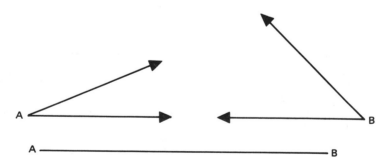

16. Given the following three segments, using the usual techniques, construct $\triangle ABC$ on your paper. Find m($\angle A$), m($\angle B$), and m($\angle C$).

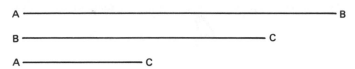

17. Given $\triangle ABC$ such that m($\angle CAB$) = 100, m($\angle ACB$) = 30, and m($\overline{AC}$) = 3, find m($\angle ABC$), m($\overline{AB}$), and m($\overline{BC}$).

18. Given $\triangle ABC$, if m($\overline{AB}$) = 5, m($\overline{AC}$) = 5, and m($\overline{BC}$) = 5, what is m($\angle ABC$), m($\angle BCA$), and m($\angle CAB$)?

19. In parallelogram $ABCD$, illustrated below, determine the measure of each indicated angle without the use of a protractor.

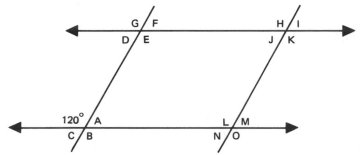

a. List the names of all angles whose measure is 120.
b. List all angles whose measure is equal to the measure of $\angle a$.
c. What is m($\angle A$) + m($\angle O$) + m($\angle H$) + m($\angle E$)?

20. On your paper draw a polygon similar to that illustrated below. Tear off three of the corners representing vertex A, B, and C. Place these corners around vertex D. Does it appear that m($\angle A$) + m($\angle B$) + m($\angle C$) + m($\angle D$) = 360, if the unit is degrees?

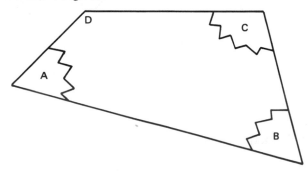

4

SIMPLE CLOSED CURVES

4.1 INTRODUCTION

Obtain a piece of paper and a sharpened pencil. Using these objects, make tracings on the flat surface of the paper with the pencil, without removing the pencil tip from the paper. Some of your drawings may look like the ones in Figure 4.1. Using this technique, we construct a model of a curve on a flat surface which is a model of a plane. It appears that every point of the curve is also a point of the plane, or that these are sets of coplanar points. Since it appears that each point of a given curve is also a point of the plane, the curves are called *plane* curves.

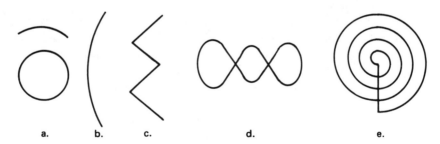

Figure 4.1

Definition 4.1—Plane Curves

For every curve s, if there is a plane which contains every point of s, then s is a *plane* curve.

Within this chapter, we shall be interested in the study of plane curves. We will study only those curves which are subsets of exactly one plane, *unless otherwise stated*.

From the above definition we see that a line is a set of points all of which belong to one plane; therefore, a line is a plane curve. Likewise, a segment is a plane curve. Still another plane curve we have studied is an angle. However, the study of lines, line segments, rays, etc. by no means exhausts the totality of plane curves; in this chapter, we shall investigate those "other" plane curves.

4.2 SIMPLE CLOSED CURVES

Some plane curves begin at one point and end at another point. There are plane curves that "start" and "stop" at the same point, thus "closing" in some portion of the plane. A "closed curve" is a plane curve that can be represented by starting and stopping at the same point. Examples of closed curves are illustrated in Figure 4.20. Each illustration may be drawn in such a manner that the curve "starts" and "stops" at A.

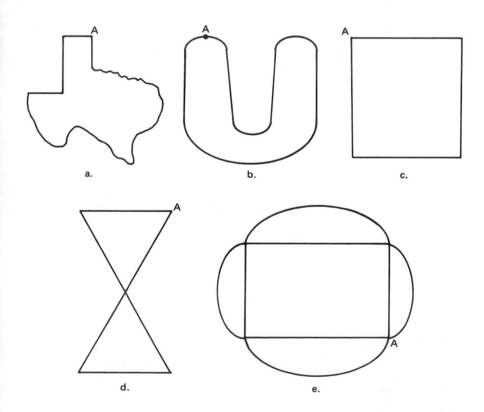

Figure 4.2

But some of these figures may be drawn so that no point (except the starting point) of the curve is "touched" twice. This is true for the figures of a., b., and c. in Figure 4.2. The reader should verify by tracing the figures on a separate sheet of paper, that in figures d. and e. at least one point must be "touched" at least twice. Figures a., b., and c. illustrate *simple closed curves*.

Definition 4.2—A Simple Closed Curve

A *simple closed curve* is a closed curve such that no point of the curve except the starting point may be "touched" twice.

Every simple closed curve separates the plane into three disjoint sets: (a) the curve, (b) the interior, and (c) the exterior. This is illustrated by Figures 4.3 and 4.4.

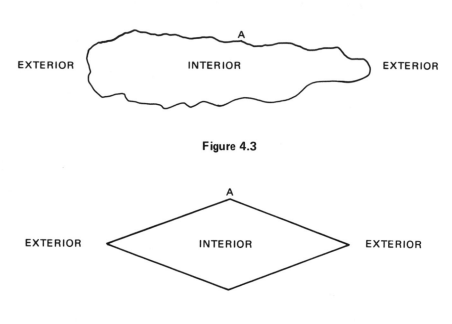

Figure 4.3

Figure 4.4

Since a line segment is also a curve, it follows that (under certain conditions), simple closed curves may be defined in terms of the union of line segments. Illustrations of these particular closed curves are given in Figure 4.5.

The simple closed curves illustrated in Figure 4.5 are called *polygons*. Each of these may be described in terms of the union of line segments. Polygon literally means "many sides."

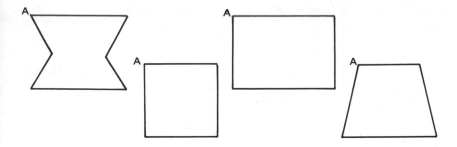

Figure 4.5

Definition 4.3—A Polygon

Let P_1, P_2, P_3, ... , P_n be n distinct points in a plane α, $(n \geqslant 3)$. Let the n segments $\overline{P_1P_2}$, $\overline{P_2P_3}$, $\overline{P_3P_4}$, ... , $\overline{P_nP_1}$ have the following properties:

a. The intersection of any two segments is one point, namely their end-points.

b. No two segments with common end-points are subsets of the same line.

Then $\overline{P_1P_2} \cup \overline{P_2P_3} \cup \overline{P_3P_4} \cup ... \overline{P_nP_1}$ is called a polygon. Any point P_1, P_2, P_3, ... , P_n is called a *vertex* of the polygon, and each of the segments $\overline{P_1P_2}$, $\overline{P_2P_3}$, $\overline{P_3P_4}$, ... , $\overline{P_nP_1}$ is called a *side* of the polygon.

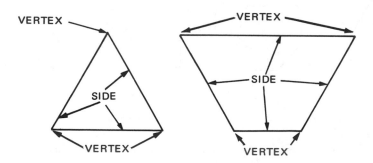

Figure 4.6

Problem and Activity Set 4.1

THINKING ABOUT PLANE CURVES

1. A frame around a picture suggests a simple closed curve. Name some other things which suggest a simple closed curve.

2. Tell which of the following are simple closed curves.

3. Tell which of the figures in Problem 2 are closed curves but not simple closed curves.
4. Do the boundaries of most states represent a simple closed curve?
5. Name one state of the United States whose boundary does not represent a simple closed curve.
6. Is the figure below a simple closed curve? Is it the union of simple closed curves? If so, how many simple closed curves?

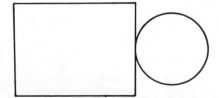

7. Mark three points on your paper like these. Draw $\overline{AC}$, $\overline{BC}$, and $\overline{AB}$. Does the figure represent a simple closed curve?

A
•

• B

C •

8. Mark four points on your paper like these. Using these points as end-points, draw four line segments so that the figure will represent a simple closed curve. Draw four other line segments so that the figure represented will *not* be a simple closed curve.

A •

• B

• C

D •

9. Draw a figure representing two simple closed curves so that their intersection is (a) exactly one point and (b) exactly two points.
10. In a map of the United States, does the union of the boundaries of California and Nevada represent a simple closed curve?
11. The curve s and the simple closed curve s_1 are shown below.

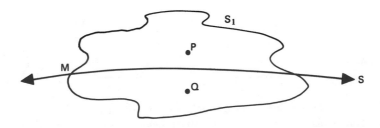

a. What is $s_1 \cap s$?
b. Shade the intersection of the interior of s_1 and the P-side of s.
12. If possible, draw two simple closed curves, A and B, so that $A \cap B$ is exactly two points; is exactly three points; is exactly four points.
13. Two simple closed curves are illustrated on page 88. Describe the region containing point P in terms of interior, exterior, and set intersection.

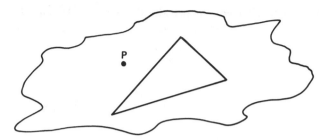

14. Draw two simple closed curves so that the intersection of their interiors is exactly three different regions.

15. Given a line ℓ and a simple closed curve C, draw a picture showing $C \cap \ell$ to be exactly three points.

THINKING ABOUT POLYGONS

16. Which of these are drawings of simple closed curves? Which are drawings of polygons?

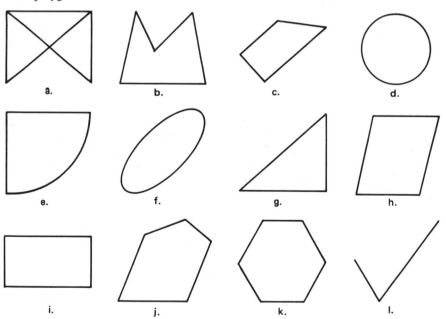

17. Draw a simple closed curve which is the union of three line segments.

18. Three line segments are illustrated below. Using the usual construction techniques, construct a polygon of three sides, with the sides congruent to the segments illustrated.

a.   b.

19. Mark four points on your paper. Label these points A, B, C, and D as indicated below. Draw $\overline{AB}$, $\overline{BC}$, $\overline{DC}$, and $\overline{DA}$. Draw $\overline{DB}$. How many polygons are illustrated in the figure? Name each side and each vertex of each polygon.

A •   • B

D •   • C

    Now draw $\overline{AC}$. Label $\overline{DB} \cap \overline{AC} = P$. How many polygons are now illustrated?

20. Answer the following "true" or "false." For each answer, draw a picture to justify your response.
   a. A polygon has the same number of vertices as it has sides.
   b. The smallest number of sides a polygon can have is two.
   c. The sides of a polygon are line segments.
   d. The sides and vertices of a polygon are subsets of the same plane.
21. Draw two examples of each of the following.
   a. Curves that are neither simple nor closed
   b. Curves that are closed but not simple
   c. Curves that are both closed and simple

4.3 CONVEX POLYGONS.

The properties of some simple closed curves may be entirely different from the properties of other simple closed curves. Of particular interest for the study of polygons is the property of *convexity*.

Definition 4.4—Convex Simple Closed Curves

    A simple closed curve, s, is said to be *convex* if for any two points C and D of the interior of s, $\overline{CD}$ is a subset of the interior of s.

Figure 4.7 illustrates simple closed curves that are *convex*. Some of the segments CD are listed.

The simple closed curves in Figure 4.8 illustrate curves that are *not* convex.

Of particular interest to the study of geometry is an investigation of convex polygons. In following sections, "polygon" shall mean "convex polygon." Some of these polygons are illustrated in Figure 4.9. The number of sides for each convex polygon is indicated below the name of the polygon.

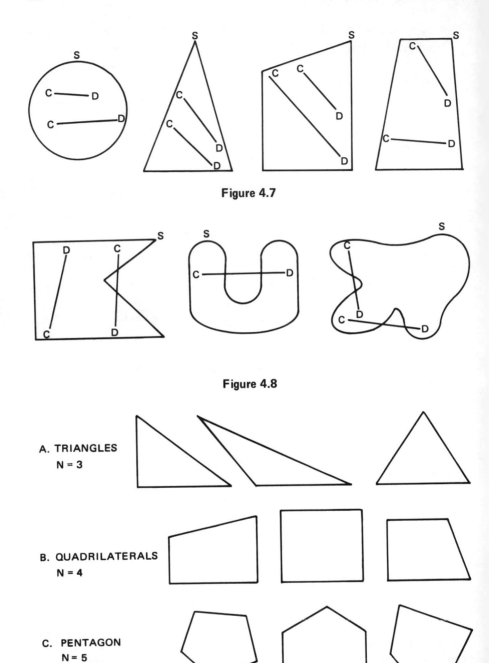

Figure 4.7

Figure 4.8

A. TRIANGLES
N = 3

B. QUADRILATERALS
N = 4

C. PENTAGON
N = 5

Figure 4.9

We may think of the *angles of the polygon*, but the angles are not a "part" of the polygon. This is so because the sides of a polygon are *segments*; the sides of an angle are *rays*. If all the angles of the polygon are congruent to each other and if all the sides are congruent to each other, then the polygon is called a *regular polygon.*

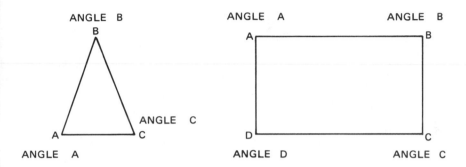

Figure 4.10

Definition 4.5—Interior of a Polygon

Let s be a convex polygon of n sides. Then the interior of s is interior of $\angle P_1 \cap$ interior of $\angle P_2 \cap$ interior of $\angle P_3 \cap ... \cap$ interior of $\angle P_n$, where each of P_1, P_2, P_3, ..., P_n is a vertex of the polygon.

Note that polygon s does not belong to this intersection. Therefore, we see that a polygon separates the plane into three disjoint subsets:
a. The set of points of the polygon
b. The set of points of the interior
c. The set of all other points of the plane, called the *exterior*
The reader should note, from the preceding, that the interior of the polygon does not belong to the polygon.

Problem and Activity Set 4.2

Tell which of the following below and on page 92 are convex simple closed curves.

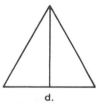

a.   b.   c.   d.

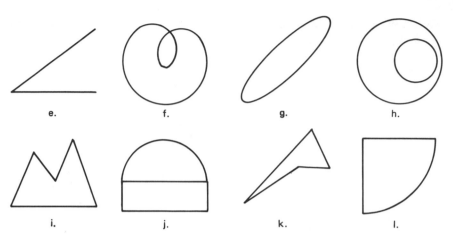

e.   f.   g.   h.

i.   j.   k.   l.

2. Using the five segments indicated below, construct a polygon which is not convex and has sides congruent to the segments.

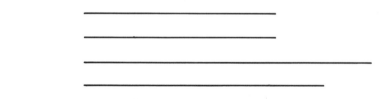

3. Using the same five segments of Problem 2, construct a convex polygon.
4. Answer the following "true" or "false." For those that are false, explain why they are false.
   a. Every complex polygon has at least three sides and there are three angles of the polygon.
   b. Any triangle is a convex polygon which has fewer sides than any other polygon.
   c. If a convex polygon has n sides, there are also n angles of the polygon.
   d. Each angle of a convex polygon is an acute angle.
   e. The set of triangles is a subset of the set of convex polygons.
   f. Each quadrilateral has four vertices.
5. Name each convex simple closed curve in the following figure.

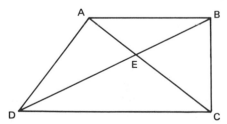

6. Draw two convex polygons whose intersection is (a) exactly one line segment, (b) exactly two points, and (c) exactly one point.
7. Given the convex polygons X and Y, below, describe the shaded region in terms of the interior and intersection of X and Y.

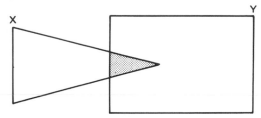

4.4 FAMILIES OF POLYGONS

Polygons of three sides and three vertices (triangles), four sides and four vertices (quadrilaterals), five sides and five vertices (pentagons), six sides and six vertices (hexagons), etc., illustrate "families" of polygons.

The first family of polygons we shall investigate is the family of triangles.

Definition 4.6—A Triangle

Let A, B, and C be non-collinear points. Then $\overline{AB} \cup \overline{BC} \cup \overline{CA}$ is the *triangle ABC*. Triangle ABC is denoted $\triangle ABC$, in which each of A, B, and C is a vertex of the triangle.

Since a triangle is also a simple closed curve, the union of the triangle, the interior, and the exterior is the plane "containing" the triangle.

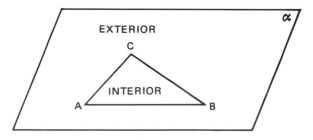

Figure 4.11

We may think of two triangles as *congruent* if one triangle can be moved onto the other one so that they "fit exactly." Remember, to say that "Figure A is congruent to Figure B," the symbol "$\simeq$" may be used. In Figure 4.12, the illustrations show that $\triangle ABC \simeq \triangle DEF$.

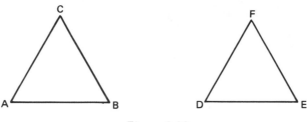

Figure 4.12

We may distinguish certain triangles from other triangles by investigating their angles and their sides. First, we may classify a triangle in terms of the *sides* of the triangle.

Definition 4.7—An Equilateral Triangle

If $\triangle ABC$ is a triangle such that $\overline{AB} \simeq \overline{BC} \simeq \overline{CA}$, then $\triangle ABC$ is called an *equilateral* triangle.

Other classifications of triangles by sides are *isosceles* triangles (at least two sides are congruent) and *scalene* triangles (no two of the three sides are congruent).

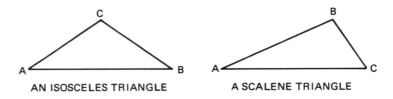

AN ISOSCELES TRIANGLE   A SCALENE TRIANGLE

Figure 4.13

Similarly, we may classify triangles according to the *angles* of the triangle.

Definition 4.8—An Equiangular Triangle

If $\triangle ABC$ is a triangle such that $\angle A \simeq \angle B \simeq \angle C$, then $\triangle ABC$ is called an *equiangular* triangle.

In Figure 4.14, $\triangle ABC$ is a triangle that is both *equilateral* and *equiangular*.

Other classifications of triangles by angles are acute, obtuse, and right. An *acute* triangle is a triangle such that each angle of the triangle is *less than* a right angle; an *obtuse* triangle is a triangle such that *one* of the angles of the triangle is *greater than* a right angle. Each category is illustrated in Figure 4.15. The reader may verify by comparing each angle of the triangles to a right angle, using the usual tracing procedures.

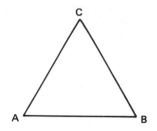

$$\overline{AB} \simeq \overline{BC} \simeq \overline{CA};\ \measuredangle A \simeq \measuredangle B \simeq \measuredangle C.$$

Figure 4.14

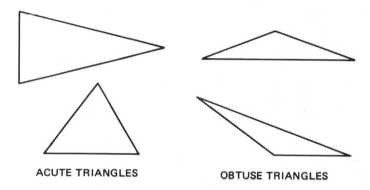

ACUTE TRIANGLES   **OBTUSE TRIANGLES**

Figure 4.15

Lastly, a *right* triangle is a triangle such that *one* of the angles of the triangle is a right angle. Each right angle is indicated by the symbol ⌐ or ⌐ as shown in Figure 4.16.

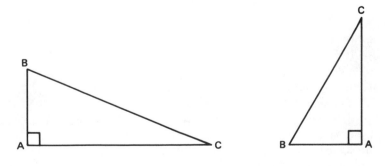

Figure 4.16

Problem and Activity Set 4.3

1. Look at the accompanying figure. Complete the following sentences, selecting your response from one of the following.
   a. a point of the triangle
   b. a point of the interior of the triangle
   c. a point of the exterior of the triangle

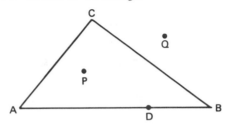

   1. Point P is _____ .
   2. Point Q is _____ .
   3. Point D is _____ .
   4. Point A is _____ .

2. By referring to the figure of Problem 1, complete the following sentences.
   a. $\overline{AB} \cup \overline{BC} \cup \overline{CA}$ is _____ .
   b. The angle with vertex A is called a(n) _____ angle.
   c. The angle with vertex C is called a(n) _____ angle.
   d. The triangle is called a(n) _____ triangle.

3. Using the drawing below, name each of the following.

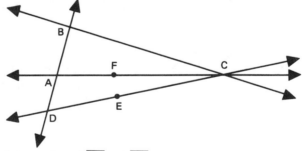

   a. The intersection of $\overline{AC}$ and $\overline{BD}$
   b. Three different triangles
   c. A segment which is not a side of a triangle
   d. A point of the interior of a triangle
   e. A point of the exterior of a triangle

4. Answer *a.-g.* "true" or "false." Write a sentence or draw a picture to justify your answer.
   The intersection of a plane and a triangle may be
   a. The empty set
   b. Exactly one point

   c. The triangle
   d. Exactly three points
   e. A line segment
   f. An angle
   g. A set which contains more than three points but not all the points of
      the triangle

5. Name the following triangles as equilateral, isosceles, or scalene.

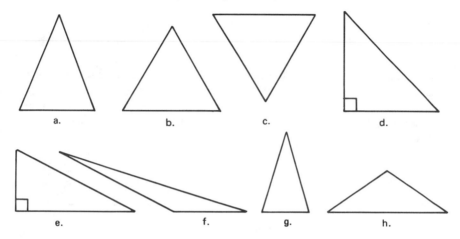

6. Name each triangle of Problem 5 as equiangular, acute, obtuse, or right.
7. Sketch the following
   a. An obtuse triangle
   b. A triangle which is both obtuse and isosceles
   c. An acute, scalene triangle
8. Three line segments are indicated below. Using these line segments, construct a triangle ABC. Classify $\triangle ABC$ according to its sides. Classify $\triangle ABC$ according to its angles.

9. Construct a triangle with one side congruent to the segment $\overline{AB}$ and with angles at each end-point congruent to $\angle A$ and $\angle B$.

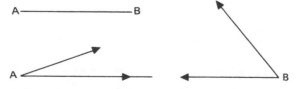

10. Construct a triangle with two sides congruent to the segments $\overline{AB}$ and $\overline{AC}$, such that the angle determined by these two sides is congruent to $\angle A$.

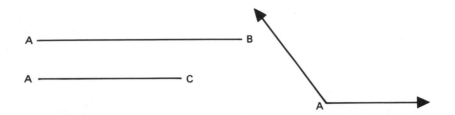

11. Suppose all sides of $\triangle ABC$ are congruent. Is $\triangle ABC$ an equiangular triangle?
12. If a triangle is isosceles, is it also equilateral? Explain your answer.
13. The figures below represent sticks of wood which can be broken at the points indicated. Select the ones that could be used to determine a triangle. Tell why you could not select the remaining.

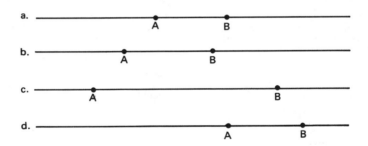

14. Name all congruent angles in the accompanying figure. $\ell_1 \parallel \ell_2$.

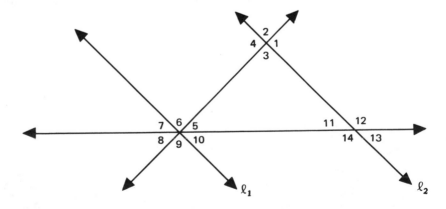

15. If $\ell_1 \parallel \ell_2$, $\ell_3 \parallel \ell_4$, and $\overline{BF} \parallel \overline{CE}$, name (a) all congruent triangles, (b) all congruent line segments, and (c) all congruent angles.

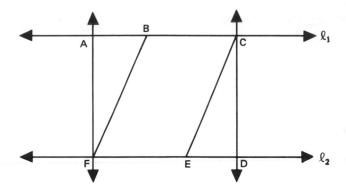

4.5 QUADRILATERALS

In prior study, we investigated the one family of polygons known as triangles. We shall now investigate another family of polygons known as quadrilaterals. In the previous section a quadrilateral was defined as a polygon with exactly four sides and four vertices. Figure 4.17 diagrams the hierarchy we have discussed so far.

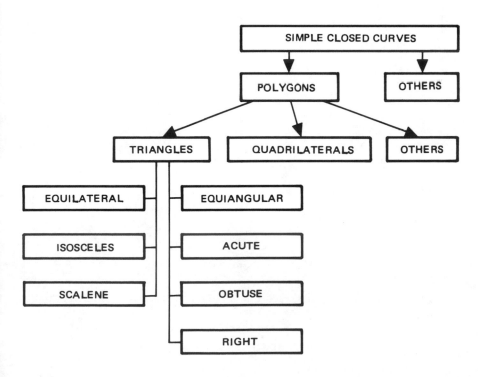

Figure 4.17

We may define a quadrilateral in the same manner as was done for a triangle.

Definition 4.9—A Quadrilateral

Let each of A, B, C, and D be coplanar points such that no three of them are collinear, $\overline{AB} \cap \overline{BC} = B$, $\overline{BC} \cap \overline{CD} = C$, $\overline{CD} \cap \overline{DA} = D$, $\overline{DA} \cap \overline{AB} = A$, and such that the intersection of any other pair of segments is the empty set. Then $\overline{AB} \cup \overline{BC} \cup \overline{CD} \cup \overline{DA}$ is a *quadrilateral*. We name the quadrilateral *ABCD*.

The order of the points in the name implies that $\overline{AB}$ is adjacent to $\overline{BC}$, $\overline{BC}$ is adjacent to $\overline{CD}$, etc. If the points were named in a different order, then different segments would be adjacent. This is illustrated in Figure 4.18 below.

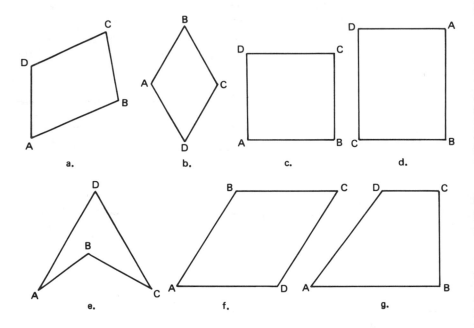

Figure 4.18

As in the case of triangles, each quadrilateral has an interior and an exterior so that:

$$\{\text{quadrilateral}\} \cup \{\text{interior}\} \cup \{\text{exterior}\} = \text{PLANE}$$

Note in Figure 4.19a. that $ABCD \cup$ {Interior $ABCD$} is a convex set, while in Figure 4.19b, $ABCD \cup$ {Interior $ABCD$} is *not* a convex set. We shall be interested in those quadrilaterals, together with their interiors, that are convex sets.

Definitions that are appropriate for the study of quadrilaterals are listed below.

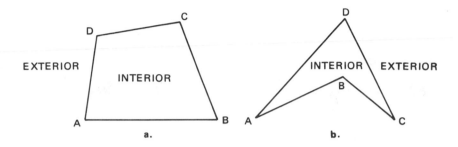

Figure 4.19

Definition 4.10—Adjacent/Opposite Sides

If $\overline{XY}$ and $\overline{YZ}$ are sides of a quadrilateral, $\overline{XY} \cap \overline{YZ} = Y$, then $\overline{XY}$ and $\overline{YZ}$ are called *adjacent sides*. Furthermore, if two sides of a quadrilateral are not adjacent sides, they are *opposite sides*.

Definition 4.11—Adjacent/Opposite Angles

If $\angle XYZ$ and $\angle WXY$ are angles of a quadrilateral, $\angle XYZ \cap \angle WXY = \overline{XY}$, then $\angle XYZ$ and $\angle WXY$ are *adjacent angles*. If two angles of a quadrilateral are not adjacent angles, they are *opposite* angles.

In summary, *opposite sides* of a quadrilateral are two sides whose intersection is the empty set; *opposite angles* are two angles whose intersection is not a side of the quadrilateral.

Definition 4.12—A Diagonal

A *diagonal* of a quadrilateral is a segment determined by two non-adjacent vertices.

There can be at most two diagonals of every quadrilateral. In Figure 4.20, $\overline{AC}$ and $\overline{BD}$ are the diagonals of $ABCD$.

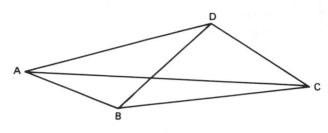

Figure 4.20

Definition 4.13—Trapezoid

If *ABCD* is a quadrilateral so that $\overline{AB} \parallel \overline{CD}$, then *ABCD* is a *trapezoid*.

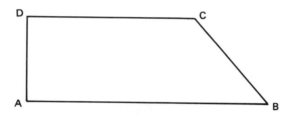

Figure 4.21

Thus a trapezoid is a quadrilateral in which only one pair of opposite sides are parallel.

Problem and Activity Set 4.4

1. Which of the figures below and on page 103 are quadrilaterals?

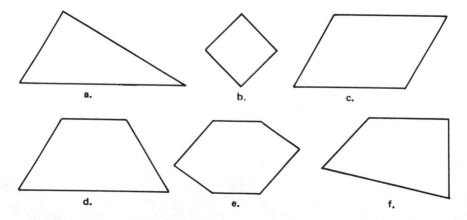

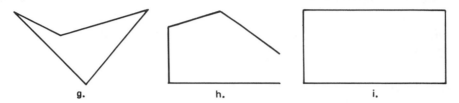

g.   h.   i.

2. Name the adjacent angles of the given quadrilateral.

Name the opposite angles of the quadrilateral.
3. Using the figure of Problem 2, name the adjacent sides of the quadrilateral; name the opposite sides of the quadrilateral.
4. Duplicate the quadrilateral of Problem 2 on your paper. Draw the two diagonals of the quadrilateral. How many polygons are now determined?
5. Draw figures similar to the ones below on your paper. If we call a "diagonal" a line segment determined by two non-adjacent vertices, draw all possible "diagonals."

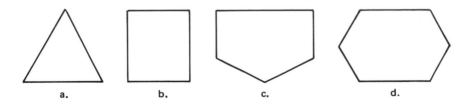

a.   b.   c.   d.

Tell how many "diagonals" were drawn for each polygon. Using this information, tell how many "diagonals" you could draw for a polygon of six sides. Tell why you selected this particular number. Draw a picture of a polygon of six sides and verify your answer.
6. Answer each of the following "true" or "false." Draw a figure to justify your answer if your selection is "false."
   a. A polygon is a quadrilateral only if it has four sides and four vertices.
   b. If a polygon does not have four sides and four vertices, then it is not a quadrilateral.
   c. Each quadrilateral has exactly four vertices.
   d. All vertices of a quadrilateral are elements of the same plane.

4.6 THE FAMILY OF QUADRILATERALS THAT ARE PARALLELOGRAMS

Of particular interest to the study of geometry are those quadrilaterals whose opposite sides are parallel.

Definition 4.14—A Parallelogram

If *ABCD* is a quadrilateral so that $\overline{AB} \parallel \overline{CD}$ and $\overline{BC} \parallel \overline{AD}$, then *ABCD* is a *parallelogram*.

Examples of parallelograms are given in Figure 4.22.

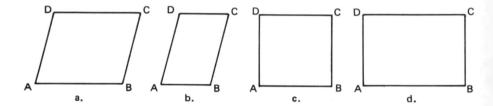

Figure 4.22

Particular characteristics of certain parallelograms distinguish them from other parallelograms. These parallelograms are the *rectangle, rhombus*, and *square*.

Definition 4.15—A Rhombus

If *ABCD* is a parallelogram and if $\overline{AB} \simeq \overline{BC} \simeq \overline{CD} \simeq \overline{DA}$, then *ABCD* is a *rhombus*.

Definition 4.16—A Rectangle

If *ABCD* is a parallelogram, and $\angle A \simeq \angle B \simeq \angle C \simeq \angle D$, then *ABCD* is a *rectangle*.

Each of Figures 4.22b. and c., is a *rhombus*. The geometric figure illustrated in c. is a particular rhombus, called a *square*.

Definition 4.17—A Square

If *ABCD* is a rhombus and if $\angle A \simeq \angle B \simeq \angle C \simeq \angle D$, then *ABCD* is called a *square*.

We are now able to extend the diagram which shows the hierarchy for simple closed curves.

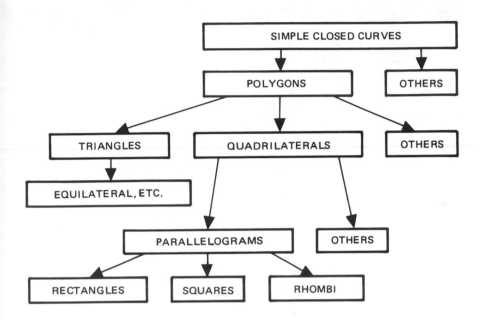

Figure 4.23

Problem and Activity Set 4.5

1. Which of the following figures represent parallelograms, assuming that segments which appear to be parallel are parallel?

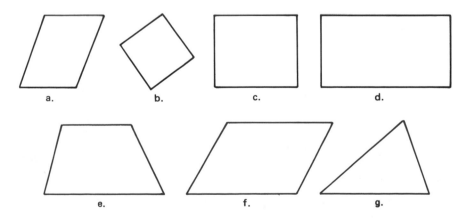

2. How many parallelograms are represented in each figure, assuming that segments which appear to be parallel are parallel?

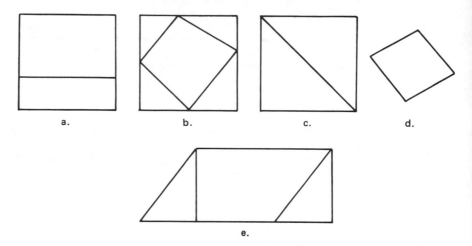

a.   b.   c.   d.

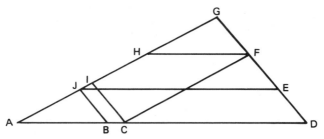

e.

3. In the accompany figure, assume that segments which appear to be parallel are parallel.

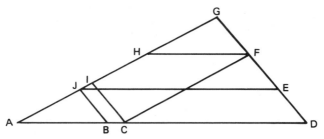

a. Name two segments congruent to $\overline{HF}$.
b. Name two segments congruent to $\overline{JB}$.
c. Name two segments parallel to $\overline{AD}$.
d. Name two segments parallel to $\overline{CD}$.
e. Name the parallelograms. (There should be 10.)
f. Name five trapezoids.

4. In the accompanying figure, assume all line segments which appear parallel are parallel.

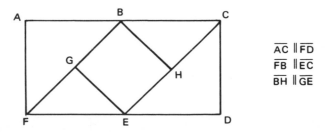

$\overline{AC} \parallel \overline{FD}$
$\overline{FB} \parallel \overline{EC}$
$\overline{BH} \parallel \overline{GE}$

a. Name each pair of congruent segments.
b. Name each pair of parallel segments.
c. Name the triangles.
d. Name the parallelograms.
e. Name the trapezoids.

5. Answer the following either "true" or "false." If your selection is "false," draw a figure or write a short sentence to justify your answer.
a. Every quadrilateral is a parallelogram.
b. Every parallelogram is a quadrilateral.
c. Every parallelogram has four congruent angles.
d. Every parallelogram has exactly two diagonals.
e. If a quadrilateral has exactly one pair of parallel opposite sides, it is a parallelogram.
f. If S = the set of squares and P = the set of parallelograms, then $S \subset P$.
g. If Q = the set of quadrilaterals and P = the set of parallelograms, then $Q \subset P$.
h. If a parallelogram determines one right angle, it follows that the parallelogram determines four right angles.
i. If a parallelogram has two adjacent, congruent sides, then all sides of the parallelogram are congruent.
j. If a parallelogram has two congruent, parallel sides, then the parallelogram is a square.
k. Opposite angles of every parallelogram are congruent.

6. Two line segments and one angle are illustrated below. Using these line segments and the angle, construct a parallelogram $ABCD$.

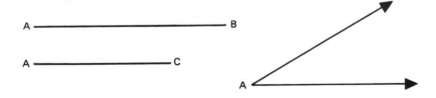

7. Using the usual techniques for constructions, draw two intersecting line segments, $\overline{AB}$ and $\overline{CD}$ (given below) so that they bisect each other. Draw segments $\overline{AC}$, $\overline{CB}$, $\overline{BD}$, $\overline{DA}$. Does the figure $ABCD$ appear to be a parallelogram? What are the diagonals?

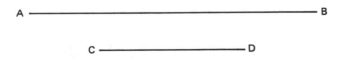

8. Using the usual techniques for constructions, draw two congruent, perpendicular bisectors of each other. Draw segments congruent to $\overline{AB}$ below. Repeat the activities of Problem 7 and answer the questions.

A ——————————————————————————— B

4.7 MEASUREMENT OF POLYGONAL REGIONS

Similar to the process for assigning measure to line segments and to angles, measure is assigned to polygonal regions, using polygonal regions as units of measure. By custom, the unit of measure is that of a "square region" which consists of a square and the interior of this square. If we let the following be a unit of measure, then by duplicating this unit eight times, we "cover up" the larger figure. Hence, the measure of the larger figure is 8.

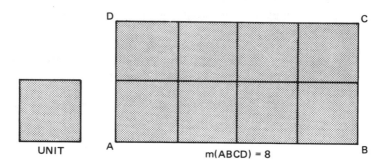

Figure 4.24

Thus, measurement of polygonal regions is again that of assigning a number to this region. It should be noted that the result does not depend on how the units are placed on the large figure, for the result will be the same in each case.

By custom, it is convenient to choose the unit region to be closely related to standard linear units. If we select the standard linear unit "inch," we may define the unit "square inch" to be a *unit square*.

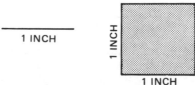

Figure 4.25

We see that if the sides of a rectangle are two inches and three inches respectively (Figure 4.26), then the rectangular closed region can be covered with exactly six unit squares. The measure of *ABCD* = 6. The unit is the square inch; and the *area* of *ABCD* is six square inches.

By observation, we see that the product of the measure (*w*) of the width (the shorter side) and the measure (ℓ) of the length (the longer side) is the measure of the area of rectangle *ABCD:*

$$m(A) = w \times \ell$$
$$= 2 \times 3$$
$$= 6$$
$$\text{Area} = 6 \text{ square inches}$$

The areas of other particular polygons may be determined in a similar manner. Consider the problem of finding the area of the parallelogram (not a rectangle) illustrated in Figure 4.27a. If we extend the sides of this parallelogram by dotted lines, as in Figure 4.27b., there are several pairs of parallel lines cut by a

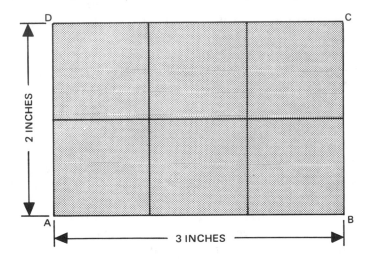

Figure 4.26

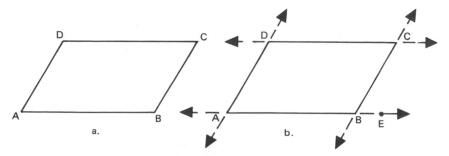

Figure 4.27

transversal. From this, we see that $\angle DAB \simeq \angle EBC$ and $\overline{AD} \simeq \overline{BC}$. From point D, draw the segment $\overline{DP}$ perpendicular to the side $\overline{AB}$. Since $\overline{AD}$ and $\overline{BC}$ are congruent, imagine the triangle ADP is moved rigidly into the position CBP'. Then P lies on the extension of $\overline{AB}$ and $\overline{AB} \simeq \overline{PP'}$. Figure 4.28 illustrates a rectangle $DPP'C$ which is the same "size" as the parallelogram $ABCD$. To find the area of the parallelogram one therefore may find the area of the rectangle.

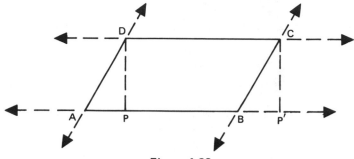

Figure 4.28

The side $\overline{AB}$ is called the *base* of the parallelogram, and the perpendicular $\overline{DP}$ is called the *altitude* of the parallelogram. Therefore, we may conclude that since $\overline{AB} \simeq \overline{PP'}$, the measure of the area of the parallelogram is "the product of the number of linear units of the base (b) and the number of linear units of the altitude (h) to this base."

$$m(A) = b \cdot h$$
$$\text{Area} = b \cdot h \text{ square units}$$

Similarly, the area of a triangle may be found by using this "rectangular" technique. Consider any triangle ABC as shown in Figure 4.29. Through C and B construct lines parallel to $\overline{AB}$ and $\overline{AC}$ respectively, meeting at some point E. Then $ABEC$ is a parallelogram. Construct the segment CD through C and perpendicular to $\overline{AB}$. The segment CD is called the altitude (h) of the triangle and $\overline{AB}$ is called the base (b) of the triangle. Notice that $\overline{CD}$ and $\overline{AB}$ are also an altitude and a base of parallelogram $ABEC$.

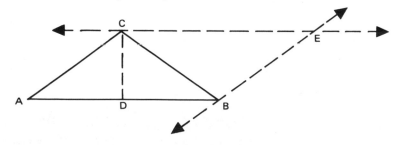

Figure 4.29

Using the usual tracing procedure, trace $\triangle ABC$ and then place this tracing so that it exactly "covers" $\triangle CBE$. This illustrates that the two triangular regions cover the whole parallelogram $ABEC$. It follows that the area of $\triangle ABC$ is one half that of the area of the parallelogram $ABEC$.

Hence, we are able to summarize: "the number of square units in the area of a triangle is one-half the product of the number of linear units in the base and the number of linear units in the altitude to this base."

$$m(A) = \tfrac{1}{2}\,b \cdot h$$
$$\text{Area} = \tfrac{1}{2}\,b \cdot h \text{ square units}$$

Problem and Activity Set 4.6

1. The following represents a polygonal region measure of "glob." Determine the measure of $ABCD$ using this unit.

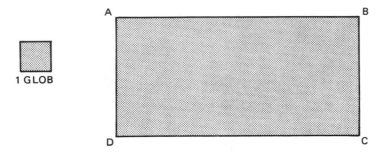

2. Determine the measure of $ABCD$ in Problem 1 if the measure is each of the units illustrated below.

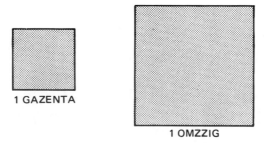

3. Each square of the grid on the right represents the polygonal unit of measure "1 gizzmo." Trace this grid on your paper. Place your grid over the figure on the left and estimate the measure of the interior of the closed curve.

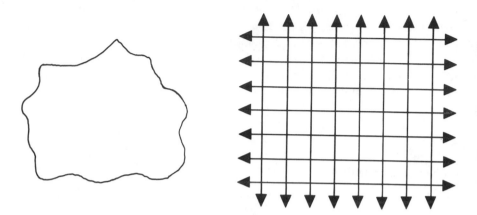

4. Find the areas of the parallelograms shown below. The dotted lines represent the altitudes of the parallelograms.

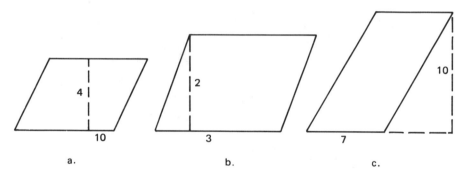

a.   b.   c.

5. Find the areas of the triangles shown below. The dotted lines represent the altitudes of the triangles.

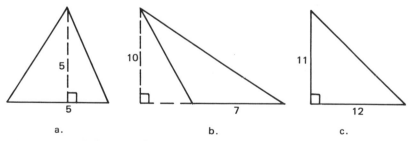

a.   b.   c.

6. In the accompanying figure, measure with an inch ruler $\overline{AB}$ and $\overline{DP}$. Find the area of $ABCD$.

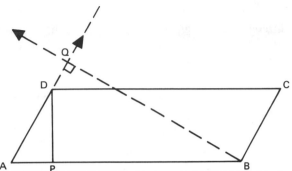

Determine the measure of $\overline{AD}$ and $\overline{BQ}$, again using an inch ruler. Find the area of *ABCD* using this base and this altitude. How do your results compare to those obtained above?

7. Use the drawing of the triangle below in order to determine the area of △*ABC*. Make all measurements to the nearest ¼ in. $\overline{AE}$, $\overline{CD}$ and $\overline{BF}$ represent altitudes.

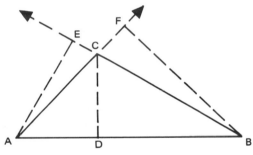

a. Determine the measure of $\overline{CD}$ and $\overline{AB}$. Find the area of △*ABC*.
b. Determine the measure of $\overline{FB}$ and $\overline{AC}$. Find the area of △*ABC*.
c. Determine the measure of $\overline{AE}$ and $\overline{CB}$. Find the area of △*ABC*.

8. The parallelogram and rectangle in the accompanying figure have bases of equal length. They also have altitudes of equal length. Trace the figures on a separate piece of paper. Cut out the figure of the parallelogram. Then cut the representation of the parallelogram into pieces that may be reassembled on the figure of the rectangle.

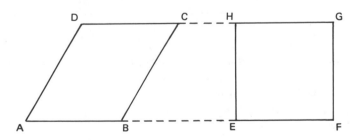

9. Three different parallelograms A, B, and C are shown. Each parallelogram has a base with measure b and an altitude with measure h. How do the three parallelograms compare in area? $\ell_1 \parallel \ell_2$.

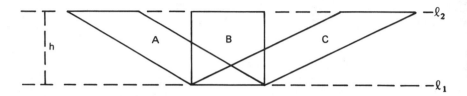

10. The figure $ABCD$ below represents a trapezoid. Trace this figure on your paper. The dotted line segment represents the altitude of the trapezoid. Draw $\overline{DB}$ on your paper, thereby constructing $\triangle DBC$ and $\triangle ADB$. Using these two triangles, find the area of $ABCD$. (NOTE: Recall that one altitude of $\triangle DBC$ is also $2''$.)

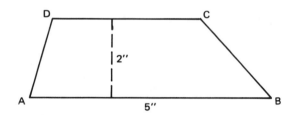

11. Using a technique similar to that of Problem 10, show that the area of trapezoid $ABCD = \frac{1}{2}h(b_1 + b_2)$ where h is the measure of the altitude; b_1 and b_2 are measures of the bases.

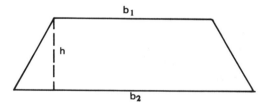

12. In order to find the area of a regular polygon, we may construct congruent triangles which are contained within the interior of the polygon by drawing line segments from the center to each vertex. The area of any such regular polygon will be the sum of the areas of the triangles. (The center of the polygon is the point P of the interior which is equally distant from the vertices of the polygon.)

Find the area of the regular polygon *ABCDEF*, below, if $m(\overline{AB}) = 4$, h is the altitude of $\triangle ABP$ and $m(h) = 5$.

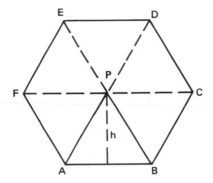

13. Using the techniques developed in this section, find the area of each polygon. Where applicable, "h" denotes the measure of the altitude.

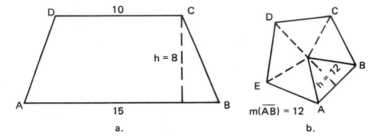

a.   b.

4.8 PERIMETER

We frequently think of the perimeter of a simple closed curve as the "length around" the curve. For example, using Figure 4.30, if we start at *A* and trace with a pencil the paths *A* to *B, B* to *C*, and *C* to *A*, we have traced a path "around" the figure. The perimeter is the length of the path from the beginning point to the ending point.

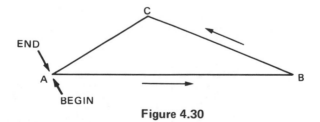

Figure 4.30

In order to determine the perimeter for any polygon, we first consider the fact that for any segment $\overline{AB}$ and any ray $\overline{DE}'$, there is a unique point B' on $\overline{DE}'$ so that $\overline{AB} \simeq \overline{DB}'$. This may be verified by using the usual tracing procedures. In Figure 4.31, $\overline{AB} \simeq \overline{DB}'$. We may say that we "duplicated" $\overline{AB}$ on the ray with D as end-point.

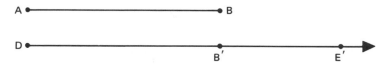

Figure 4.31

With this in mind, we can duplicate each of the segments of any polygon on some ray in such a manner that the segments are "side-by-side" and have exactly one point in common. In Figure 4.32, we see that $\overline{AB} \simeq \overline{DB}'$, $\overline{BC} \simeq \overline{B'C}'$, and $\overline{CA} \simeq \overline{C'A}'$. Using the usual processes for measurement, we can now determine the measure of $\overline{DA}'$. *If this measure is "n," then the perimeter of* $\triangle ABC = n$ *linear units.*

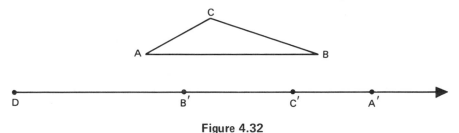

Figure 4.32

Frequently, the lengths of each segment of a polygon are given as "2 feet," or "5 inches," or in terms of some other convenient unit. This is illustrated in Figure 4.33. If we duplicate these segments side-by-side on some ray with end-point D so that any two segments have exactly one point in common, we find the measure of $\overline{DA}'$ to be 9. The *length* of $\overline{DA}'$ is 9 inches, which is also the perimeter of $\triangle ABC$. Investigate Figures 4.33 below, and 4.34 on page 117.

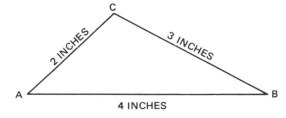

Figure 4.33

Figure 4.34

In general, if the measures of the three sides of any triangle ABC are given as m, n, and p, then:

$$\text{Perimeter } \triangle ABC = (m+n+p) \text{ units}$$

Two of the most familiar closed curves are the *rectangle* and *square*. Each is a four sided polygon (in a plane) which has a right angle at each of its corners. If the "length of the rectangle" and the "width of the rectangle" are denoted ℓ linear units and w linear units respectively, we may use the usual duplicating processes in order to determine the perimeter of the rectangle.

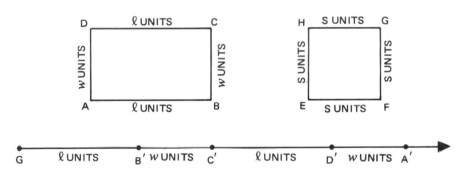

Figure 4.35

If the measure of the four sides of any rectangle is given as ℓ, w, ℓ, and w, then:

$$\text{Perimeter of rectangle } ABCD = (\ell + w + \ell + w) \text{ units}$$
$$= (2\ell + 2w) \text{ units}$$

In the above, if $\ell = 3$, $w = 2$ and the unit of measure is the inch, then:

$$\text{Perimeter of rectangle } ABCD = (2 \cdot 3 + 2 \cdot 2) \text{ inches}$$
$$= 10 \text{ inches}$$

Since a square is a particular rectangle, the same procedures may be used to determine the perimeter. If the measure of each side of any square $CDEF$ is s, then:

$$\text{Perimeter of square } CDEF = (s + s + s + s) \text{ units}$$
$$= 4 \cdot s \text{ units}$$

By using the ideas of duplication, the perimeter of *any* n-sided polygon may be determined. If the measure of the *n* sides of a polygon is s_1, s_2, s_3, ... s_n, then the perimeter is:

$$(s_1 + s_2 + s_3 + ... + s_n) \text{ units}$$

4.9 SUMS OF ANGLES

As in the case for finding the perimeters of different polygons, we may also think of "adding angles." In order to do this, we must be able to "duplicate" angles, a process very similar to that of duplicating line segments.

Consider the fact that for any angle ABC and any ray $\overrightarrow{DE}$, there is a unique ray $\overrightarrow{DF}$ (in one of the half-planes determined by $\overrightarrow{DE}$) so that $\angle ABC \simeq \angle EDF$.

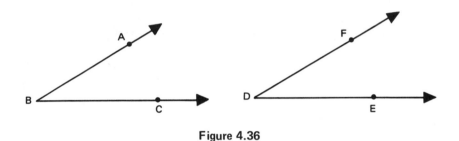

Figure 4.36

In Figure 4.36 $\angle ABC \simeq \angle EDF$. This may be verified using the usual tracing procedures.

With this in mind, we can duplicate in the same plane a given angle "on top of" another given angle in such a manner that the angles are "side-by-side" and have exactly one ray in common. In Figure 4.37, we see that $\angle ABC \simeq \angle DEA'$ and $\angle DEA' \cap \angle DEF = \overrightarrow{DE}$.

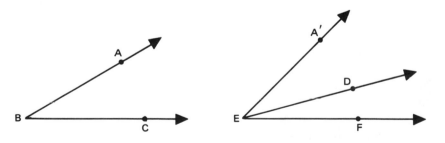

Figure 4.37

Using the usual processes for measurement of angles, we can determine the measure of $\angle FEA'$. If $m(\angle DEF) = n$ and $m(\angle ABC) = m$, then by using the duplication technique, we observe that $m(\angle FEA') = (m + n)$.

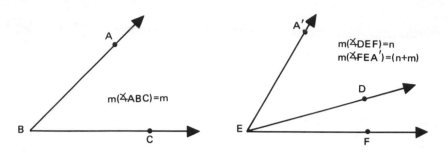

Figure 4.38

In Figure 4.38, if $m(\angle ABC) = 50, m(\angle DEF) = 20$, then $m(\angle FEA') = 70$. The reader should verify this using a protractor.

Hence, we have devised a method to "add angles." For any $\angle ABC$ and $\angle DEF$, if $m(\angle ABC) = n$ and $m(\angle DEF) = m$,

$$m(\angle ABC) + m(\angle DEF) = (m + n)$$

Recall that one standard unit of measure for angles is the degree. We may then say that the sum of the measures of $\angle ABC$ and $\angle DEF$ is $(n+m)$ degrees. In Figure 4.38,

$$(m(\angle ABC) + m(\angle DEF)) = (50 + 20)° = 70°$$

Problem and Activity Set 4.7

1. Find the perimeter of each of the following. Line segments that appear parallel are parallel; line segments that appear congruent are congruent.

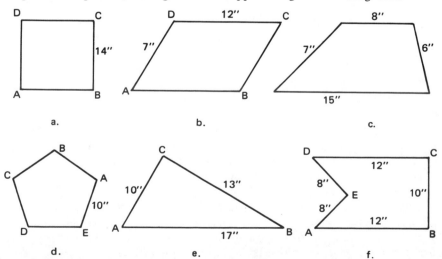

2. Using a protractor, determine the sum of the measures of the angles of the following figures.

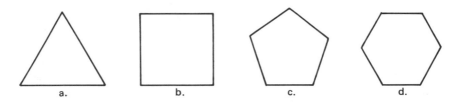

   a.   b.   c.   d.

3. By observing the data of Problem 2, what do you think will be the sum of the measures of the angles of polygon with seven sides? A polygon of eight sides?

4.10 MEASURE OF RECTANGULAR PRISMS

A figure that looks like a cardboard box is called a *rectangular prism*. This is illustrated in Figure 4.39.

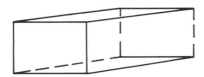

Figure 4.39

A prism is one of the most familiar of all geometric figures, and you can find many examples of a prism. The walls of your classroom may represent the points of a prism.

The "flat sides" of the prism are called *lateral faces*. Each of the faces is a closed rectangular region and is therefore a subset of some plane. You probably have noticed that there are six such faces of the prism. A prism is the union of the set of all points of the six faces.

A prism separates space into three distinct subsets of points: the prism, the points of the interior, and the points of the exterior. If you think of your classroom as a prism, you and your fellow students could represent points of the interior; the students outside the building could represent points of the exterior.

Recall that the intersection of two planes is either a line or the empty set. For each face of a prism, there is *exactly one other* face such that the intersection of the planes of these faces is the empty set. These faces are called *opposite faces*.

Note that for each prism, there are three pairs of opposite faces. Since the planes of opposite faces are parallel planes, they "cut" each other in such a manner that opposite faces are also congruent.

The intersection of two faces which are not opposite is a line segment. These segments are called *edges* of the prism. The point which is the intersection of three faces is a *vertex* of the prism.

The measure of any pair of parallel edges is the same; hence, there can be at most three different measures for the edges of a prism. In Figure 4.40, the number of linear units in the lengths of three edges are marked ℓ, w, and h, since these edges are often called the *length, width*, and *height* of the prism.

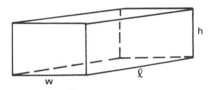

Figure 4.40

Since all the faces are rectangular closed regions, the lateral surface area of each face can be determined using usual techniques of measurement. The sum of the areas of the measures of the six lateral surfaces is called the *lateral surface area*.

$$\text{Lateral Surface Area} = (\ell w + \ell w + \ell h + \ell h + wh + wh) \text{ square units}$$
$$= (2\ell w + 2\ell h + 2wh) \text{ square units}$$
$$= 2(\ell w + \ell h + wh) \text{ square units}$$

The expression *rectangular solid* refers to the set of points which is the union of the prism and its interior. To determine the measure of a rectangular solid is to determine its *volume*. The expression "volume of the interior of a rectangular prism" can be used in the same manner as the expression "area of the interior of a rectangle" was used. As in the case for assigning a number to line segments and to polygonal regions, a number is assigned to rectangular solids using other rectangular solids as units of measure. This number is the measure of the *volume*.

By custom, the unit of measure is the "cube," which is a rectangular solid with congruent edges. If we let the following be a unit of measure, then by duplicating this unit 16 times, we "fill up" the larger figure. Hence, the measure of the larger figure is 16.

By custom, it is convenient to choose the rectangular solid to be closely related to standard linear units and to standard region units. If we select the standard linear unit "inch," we may define the "cubic inch" to be a unit solid.

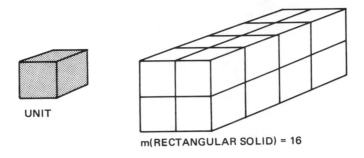

UNIT

m(RECTANGULAR SOLID) = 16

Figure 4.41

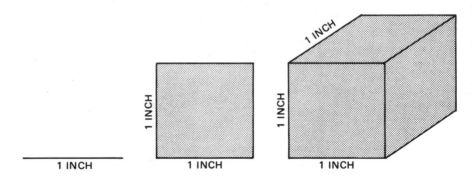

1 INCH   1 INCH   1 INCH

Figure 4.42

We see that if the edges of a rectangular solid are two inches, three inches, and four inches respectively, then the solid can be "filled up" with exactly 24 unit cubes. The measure of the rectangular solid is 24. The unit is the cubic inch, and the *volume* is 24 cubic inches.

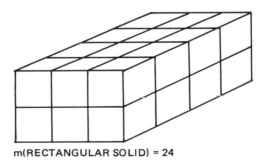

m(RECTANGULAR SOLID) = 24

Figure 4.43

By observation, we see that the product of the measures of the width (w), length (ℓ), and height (h) is the measure of the volume of the rectangular solid:

$$m(V) = w \times \ell \times h$$
$$= 3 \times 4 \times 2$$
$$\text{Volume} = 24 \text{ cubic inches}$$

Since we think of the product ($w \times \ell$) as the measure of the area of the base, we may think of the volume of *any* prism as

$$m(V) = m(\text{area of base}) \times h$$

where h is the height of the prism.

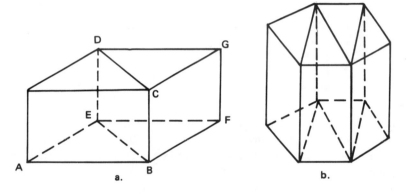

Figure 4.44

Consider the triangular prism *ABCDE* in Figure 4.44a. Since the area of $\triangle ABE$ is one-half that of square *ABFE*, it follows that the volume of *ABCDE* is one-half that of the rectangular solid.

Similarly, we may determine the volume of any prism if we "cut up" the prism into triangular prisms as illustrated in Figure 4.44b. The base of the prism has six sides. The prism itself is "cut up" into four triangular prisms.

Problem and Activity Set 4.8

1. Find (*a*) the number of cubic units of volume and (*b*) the number of square units of lateral surface area for each of the prisms shown on page 124.

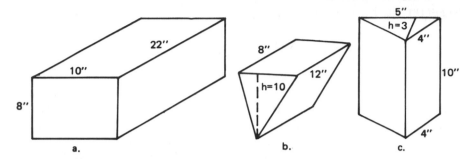

a.   b.   c.

2. Find the number of cubic units of volume for each prism shown below:

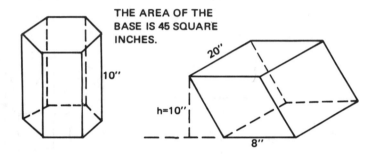

THE AREA OF THE
BASE IS 45 SQUARE
INCHES.

3. A container in the shape of a prism is 20 inches high and the volume is 3300 cubic inches. How many square inches are there in the base?

4. A triangular prism has a base which represents a right triangle with the two perpendicular sides three inches and five inches in length. If the prism is 25 inches high, what is the volume in cubic inches?

5. Mathematicians have determined that the volume of a pyramid is one-third that of a prism whose base is congruent to the base of the pyramid and whose height is the same as that of the pyramid. The formula $V = \frac{1}{3} Bh$, where B stands for the number of square units of area in the base and h the number of linear units in the height, may be used for finding the volume of any pyramid. Find the volume of each of the following.

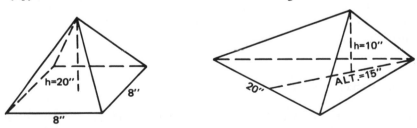

6. What is the height of a pyramid whose volume is 3375 cubic inches and whose base is a square, 15 inches on a side?

7. The side of the square base of a pyramid is doubled. The height of the pyramid is halved. How is the volume affected?

8. Find the volume of each of the following if the area of each base is 144 square inches and each height is 20 inches.

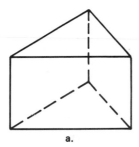
a.

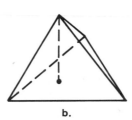
b.

9. Find the volume of (a) a triangular prism whose height is 40 in. and (b) a triangular pyramid whose height is 40 ft, if the *base of each* is illustrated below.

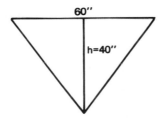

5

CIRCLES AND SPHERES

5.1 INTRODUCTION

In this chapter, we will study other sets of points of a plane and of space. The particular sets of points we will investigate are called *circles* and *spheres*. Within this section we will study circles; in following sections we will study spheres.

Definition 5.1—A Circle

Given a point P and a segment $\overline{AB}$, the *set* of all points, X, in a plane such that $\overline{PX} \simeq \overline{AB}$, is a *circle*.

The point P in Figure 5.1 is called the *center* of the circle and the line segments $\overline{PQ}$, $\overline{PR}$, $\overline{PS}$, and $\overline{PT}$ are called *radii* of the circle. We may refer to this circle as "circle P."

A circle is a simple closed curve in a plane and hence has an exterior and an interior. In Figure 5.2, point A is a point of the interior of the circle and point B is a point of the exterior. The center of the circle is point P, and one radius of the circle is labeled $\overline{PC}$.

We frequently denote the length of the radius as r linear units. Thus, the circle with center P and radius r linear units is the set of all points, X, in a plane,

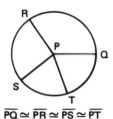

$$\overline{PQ} \simeq \overline{PR} \simeq \overline{PS} \simeq \overline{PT}$$

Figure 5.1

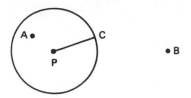

Figure 5.2

$m(\overline{PX}) = r$. The exterior is the set of all points Y such that $m(\overline{PY}) > r$ and the interior is the set of all points Z such that $m(\overline{PZ}) < r$. Recall r is a number and $m(\overline{PZ})$ is a number.

The interior of a simple closed curve and the curve itself is called a region. In Figure 5.3, the interior of the circle is shaded. The union of a circle and its interior is called a *circular region*.

Two or more distinct circles that lie in the same plane and have the same point P as center are called *concentric circles*.

In Figure 5.4, we see that $\overline{PQ} \neq \overline{PQ'}$, but each circle has as center point P, and both circles are in the same plane.

Figure 5.3

Figure 5.4

Problem and Activity Set 5.1

1. Given the circle illustrated below.

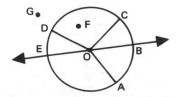

a. Name two radii of the circle.

b. Is point F an element of the interior or exterior of the circle?

c. Name seven points that are members of the circular region.

d. Is $\overline{OC} \simeq \overline{OB} \simeq \overline{OE}$?

2. Let C be a circle with center P and radius r units. Let S be any other point in the plane. Answer the following and justify your answer by drawing a figure.

a. How many points are elements of the set $C \cap \overrightarrow{PS}$?

b. How many points are elements of the set $C \cap \overleftrightarrow{PS}$?

c. How many points are elements of the set $C \cap \overline{PS}$? Does your answer depend on the choice of S? If so, why?

3. Choose two distinct points and label them P and Q. Draw two circles with center at P such that Q is a point of the interior of one circle and a point of the exterior of the other. Label the circles A and B.

4. Choose two distinct points and label them P and Q. Draw a circle with center P and radius $\overline{PQ}$. Draw another circle with center Q and radius $\overline{PQ}$. Call the circles C and D.

a. How many points are in the set $C \cap D$?

b. Let R denote the interior points of C, and R_1 denote the interior points of D. Shade $R \cap R_1$.

c. Make another copy of the two circles and shade the intersection of the interior of D and the exterior of C.

d. Make another copy of the two circles and shade the union of the interiors of the two circles.

5. The two circles, A and B, shown below, are subsets of the same plane and have center P.

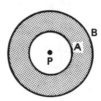

a. What is the set $A \cap B$?

b. Give a word description of the shaded region, using terms such as "intersection," "exterior," and "interior."

6. In the accompanying figure, let the half-plane on the B-side of $\overleftrightarrow{AC}$ be called H. Let the half-plane on the D-side of $\overleftrightarrow{AC}$ be called K.

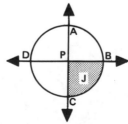

a. Draw a picture and shade $H \cap K$.
b. What is the intersection of H, J, and the circle?
c. Using terms such as "half-plane" and "intersection," describe the shaded portion of the figure represented by J.

7. In the accompanying figure, the center of each circle is a point of the other circle. Copy this figure on your paper and shade the union of the exteriors of the two circles.

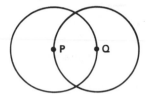

8. Denote the points A, B, and C on your paper as illustrated below.

• C

A •

• B

a. Draw $\overline{AB}$.
b. Draw the circle with center A so that B is a point of the circle. Call this circle A.
c. Draw a circle with center C with radius $\overline{BC}$. Call this circle B.
d. What is $\overline{AB} \cap$ circle B?
e. What is circle $A \cap$ circle B?
f. What is $\overline{AB} \cap$ interior circle B?
g. What is circle $A \cap$ interior circle B?

5.2 CHORDS, DIAMETERS, AND TANGENTS

Given a circle with center point P and a line ℓ in a plane. Exactly one of the following is true.

a. Circle $P \cap \ell = \{ \ \}$.

Figure 5.5

b. Circle $P \cap \ell$ is exactly one point.

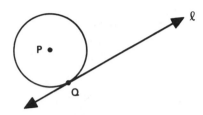

Figure 5.6

c. Circle $P \cap \ell$ is exactly two points.

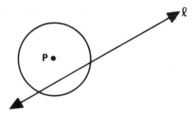

Figure 5.7

Of particular interest to the reader are cases *b* and *c*.

Definition 5.2—A Tangent Line

Given a circle with center P and a line ℓ in a plane, if circle $P \cap \ell$ is exactly one point, the line ℓ is called a *tangent* to the circle. If circle $P \cap \ell$ = point Q, then Q is the *point of tangency*. See Figure 5.8.

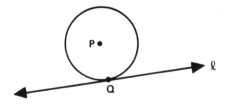

Figure 5.8

If circle $P \cap \ell$ is exactly two points, the line ℓ is called a *secant* line. In Figure 5.9, the two points of the intersection are B and Q.

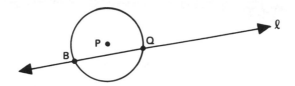

Figure 5.9

The reader should note that the intersection of circle P and line ℓ determines the line segment $\overline{BQ}$ and ℓ. This line segment is called a *chord*.

Definition 5.3—A Chord and a Diameter of a Circle

Given a secant line ℓ of circle P, the line segment determined by the two points of intersection of circle P and ℓ is called a *chord* of the circle. If this chord contains the center, the chord is called a *diameter* of the circle.

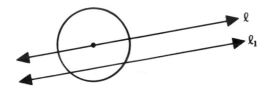

Figure 5.10

We may intuitively think of a diameter as the union of two distinct radii which are line segments of the same line. In Figure 5.11, we see that $\overline{SP} \cup \overline{PS'} \in \overleftrightarrow{SS'}$; $\overline{PQ} \cup \overline{PQ} \in \overleftrightarrow{QQ'}$, etc.

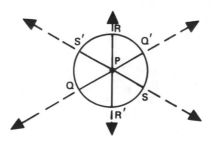

Figure 5.11

Using any linear unit, if r and d are measures of the radius and diameter, respectively, we see that:

$$d = r + r$$
$$= 2r,$$

and

$$r = \tfrac{1}{2}d.$$

Problem and Activity Set 5.2

1. How many tangent lines are represented in each figure below?

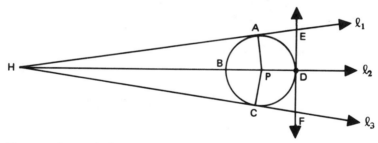

2. In the drawing below, how many lines are tangent to circle P?

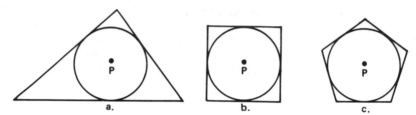

    a. Name each point of tangency.
    b. Name one chord determined by ℓ_2.
    c. Name four radii of the circle.
    d. Name a secant line.

3. Given a circle P and a point A which is a member of the exterior of the circle, a tangent to the circle containing point A can be constructed by following the steps labeled 1, 2, 3, and 4 in the accompanying illustration. Point D is the point of tangency.

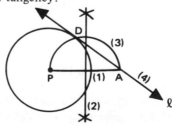

Copy circle P below. Construct *two* tangent lines, ℓ_1 and ℓ_2 so that $A \in \ell_1$, and $A \in \ell_2$. Label the points of tangency D and F, respectively.

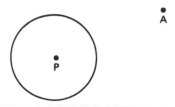

4. Name the following as either (*a*) a chord, (*b*) a diameter, (*c*) a radius, (*d*) a secant line, or (*e*) a tangent line.

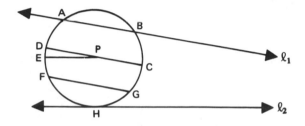

5. It is possible to construct a tangent to a given circle at a given point of the circle. This is illustrated below, where A is a point of the circle P. The line ℓ is tangent to circle P at point A. What is the $m(\angle PAD)$?

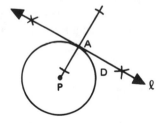

Copy the circle 0, below, and construct a tangent to the circle at the points Q and R. Label these tangent lines ℓ_1 and ℓ_2. Label $\ell_1 \cap \ell_2$ point B.

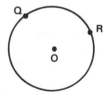

6. Copy the circle P, on page 134, on your paper. Draw chord $\overline{AB}$ of circle P. Construct a line which is the perpendicular bisector of $\overline{AB}$.

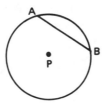

Does the line which is the perpendicular bisector of $\overline{AB}$ contain P as an element?

7. Given a circle P and points A and B of $\overline{AB}$, which is a diameter of circle P. Construct a line ℓ_1 tangent to P at A; construct another line ℓ_2 tangent to P at B. Does it appear that $\ell_1 \parallel \ell_2$? What is the measure of each of the angles with A as a vertex? With B as a vertex?

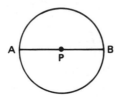

8. Answer the following "true" or "false."
   a. The center of a circle is a part of the circle.
   b. A circular region does not include the circle.
   c. A secant of a circle can be a chord of the circle.
   d. A diameter of a circle is a chord of the circle.
   e. Concentric circles are also congruent circles.
   f. Let A be a circle with center P and B be a circle with center P. Then $A \cap B$ is exactly one point.

5.3 ARCS OF CIRCLES

Recall that a point on a line separates the line into two half-lines. But this is not true for simple closed curves. In each of the diagrams of Figure 5.12, if the reader will trace a path starting at A and moving in a clockwise path, the path will return to A. This will also be true if the reader traces a counter-clockwise path.

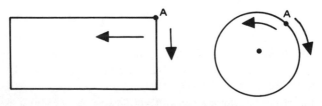

Figure 5.12

In order to separate a circle into two parts, label two distinct points of the circle A and B.

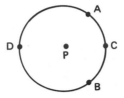

Figure 5.13

One part of the circle contains the point C and the other part of the circle contains the point D. No path from A to B along the circle can avoid at least one of the points, C or D. Thus, it takes two different points of the circle to separate the circle into two distinct parts.

The point C is said to be "between" points A and B. Likewise, the point D is also "between" A and B. Hence, the idea of "betweenness" for points of lines and the idea of separation for lines do *not* parallel those ideas for circles. It takes two points to separate a circle into two parts. Also, two distinct points of two separate parts of the circle are "between" the points that separate the circle.

A part of a circle is called an *arc*. In Figure 5.14, the points A and B separate the circle into two arcs. Each of these arcs contains A and B as end-points. The symbol for arc is ⌒, and to denote which one of the two arcs of a circle is discussed it becomes necessary to include a point between A and B. In Figure 5.14, arc ADB ($\overarc{ADB}$) is one arc of the circle; $\overarc{ACB}$ is another arc of the circle.

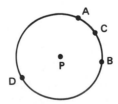

Figure 5.14

Arcs have some properties that parallel those for line segments. For example, there is a one-to-one correspondence between the points of $\overarc{ACB}$ and $\overline{DE}$.

Also, as a point such as F in Figure 5.15, separates a segment into two segments, so the point C separates $\overarc{ACB}$ into two arcs. Finally, as a line segment has end-points, so an arc has end-points.

The end-points of a diameter determine two "special" arcs called *semi-circles*. In Figure 5.16, each of $\overarc{ACB}$ and $\overarc{ADB}$ is a semi-circle.

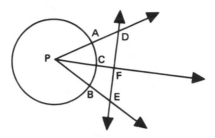

Figure 5.15

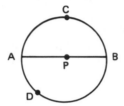

Figure 5.16

The end-points of a semi-circle and the center of a circle are points of a straight line. For all other arcs this is not true. In Figure 5.17, the point P is not a point of the straight line containing the end-points of $\widehat{ACB}$.

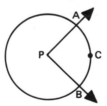

Figure 5.17

The union of the two rays $\overrightarrow{PB}$ and $\overrightarrow{PA}$ is the angle APB. Such an angle is called a *central angle*. A central angle is an angle whose vertex is the center of the circle. The measure of a central angle is determined by using the usual processes.

To assign measure to an arc, one standard unit of measure is *one degree of arc*. If we think of a circle separated into 360 congruent arcs, each arc determines one unit of arc measure. In order to determine the measure of an arc, we may use the ideas of duplicating this unit "side-by-side" on the circle so that consecutive arcs have exactly one point in common. In Figure 5.18, the measure of $\widehat{ACB} = 70$. Only the multiples of ten degrees of arc are shown.

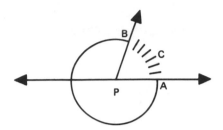

Figure 5.18

The reader should note that the union of the two rays $\overrightarrow{PA}$ and $\overrightarrow{PB}$ in Figure 5.18 also is a central angle whose measure in degrees is 70. Using these two units, we see that one degree of arc is determined by a unit angle of one degree. In Figure 5.18, $m(\overparen{ACB})$ in degrees of arc is 70, and $m(\angle APB)$ in degrees is also 70. The reader may verify this using the usual techniques for measurement.

The reader should note that arc measure is not a measure of length. In Figure 5.19, the two arcs $\overparen{ACB}$ and $\overparen{DEF}$ have the same central angle, and hence the same arc measure. If $m(\angle DPB) = 70$, then $m(\overparen{DEF}) = m(\overparen{ACB}) = 70$.

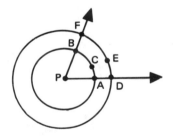

Figure 5.19

It appears that $\overparen{ACB}$ is "less than" $\overparen{DEF}$, although their arc degree measures are the same. Hence, two arcs may have the same arc degree measure but not be congruent. A more extensive discussion of arc length will follow in a later section.

Problem and Activity Set 5.3

1. Using the drawing on page 138, identify the "shortest" arc containing the following points, where the points are *not* end-points of the arc.

   a. *D*   b. *F*
   c. *A*   d. *B*
   e. *C*   f. *E*

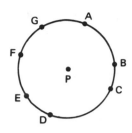

2. Refer to the illustration of Problem 1. Name three different arcs with end-point A.

3. Using the illustration of Problem 1, name the point (or points) included in each of the following arcs, which are *not* end-points of each of the following arcs.
   a. $\overparen{BCD}$ b. $\overparen{ACD}$
   c. $\overparen{ADF}$ d. $\overparen{DCG}$
   e. $\overparen{DEB}$ f. $\overparen{EBD}$

4. Using the illustration of Problem 1, point D separates $\overparen{FCB}$ into two arcs. Name these arcs.

5. In the figure below, determine with the use of your protractor the measure of the following arcs. Indicate your results with correct use of symbols, for example $m(\overparen{ABC}) = 90$.

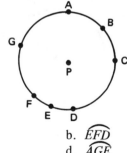

   a. $\overparen{AED}$ b. $\overparen{EFD}$
   c. $\overparen{BCD}$ d. $\overparen{AGE}$
   e. $\overparen{EDC}$ f. $\overparen{ABC}$

6. On your paper, draw a circle with center P. Select a point of the circle and label it A. Mark off and label arcs with the following measures. (Is it possible to construct several different figures?)
   a. $m(\overparen{AB}) = 20$ b. $m(\overparen{ABC}) = 45$
   c. $m(\overparen{BCD}) = 60$ d. $m(\overparen{CDE}) = 70$

7. Draw a circle with radius length of three inches. Starting at any point of the circle, use a compass with the same setting (three inches) to describe a series of "equally spaced" marks around the circle.
   a. How many congruent arcs were determined?
   b. Did you get back to *exactly* the same point from which you started?
   c. What is the measure of the central angle determined by one of the congruent arcs?

8. Refer to the arc *ABCDEF* shown in the drawing.

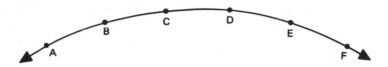

Determine the following.

a. $\widehat{AC} \cup \widehat{BD}$ b. $\widehat{AC} \cap \widehat{BD}$

c. $\widehat{AD} \cup \widehat{DF}$ d. $\widehat{AD} \cap \widehat{DF}$

9. Demonstrate a one-to-one correspondence between the sets of points of the two semi-circles of a given circle which are determined by a diameter.

10. Using the figure below, determine $m(\widehat{ACB})$ using your protractor. $\overline{AD}$ is a diameter of the circle.

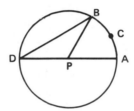

What is $m(\angle ADB)$? What is $m(\angle APB)$? How does $m(\angle ADP)$ compare to $m(\angle APB)$?

11. Using the figure below, determine $m(\angle ABC)$. $\overline{AC}$ is a diameter of the circle. Next, determine $m(\angle AB'C)$. How does $m(\angle ABC)$ compare to $m(\angle AB'C)$?

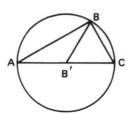

12. Draw a circle with center P and diameter 3 units. Draw chord $\overline{AB}$ so that $m(\overline{AB}) = 1$. Determine $m(\angle APB)$. Example:

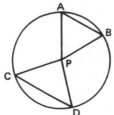

Draw another chord $\overline{CD}$ of circle P so that $m(\overline{CD}) = 2$. Determine $m(\angle CPD)$. How does $m(\angle APB)$ compare to $m(\angle CPD)$?

5.4 CIRCUMFERENCE AND AREA OF CIRCLES

The processes for determining the measure of the circumference of a circle are complicated ones. The *circumference* of a circle is the length of the closed curve defined as a circle. We determine the measure of the circumference by using linear units of measure. For example, we may think of the set of points on the rim of a wheel as depicting the points of a circle with the center point located at the axle of the wheel. If we place a string on the rim of the wheel, starting at some point A, and tightly wrap this string on the rim until we "return" to point A, we may then measure this string, using linear units, in order to obtain the measure of the circumference of the circle. In Figure 5.20, assume that we cut the string at point B, which coincides with point A.

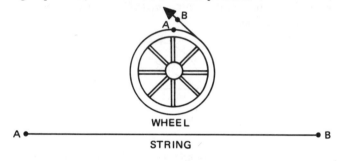

Figure 5.20

Experimentation with similar physical objects reveals that the constant ratio of the measure of the circumference of the circle and the measure of the diameter of the circle is *approximately* 3.14. Mathematicians have proved that this number is irrational, hence we cannot state an exact decimal or fractional equivalent for this ratio. The special symbol π represents this ratio. If we let c be the measure of the circumference and d be the measure of the diameter of a circle, then:

$$\frac{c}{d} = \pi \ \text{ or } \ c = \pi(d)$$

Similar problems occur when attempting to find the measure of the *circular region* determined by the circle. In order to determine this measure, we resort to the use of unit squares as the unit of measure.

In Figure 5.21, the point P is the center of the circle and also the center of the square *BDEF*. The segments $\overline{PA}$ and $\overline{PC}$ are radii of the circle with measure r. Angle *CPA* is a right angle. The measure of the square *DBFE* is four times the measure of the square *CBAP*.

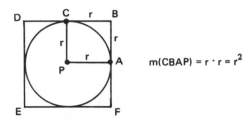

Figure 5.21

Note that the measure of the interior region of the circle is less than the measure of the interior region of *BDEF*. Mathematicians have determined that the measure of the interior region determined by the circle is approximately three times the measure of *CBAP*.

The measure of the square *CBAP* is $r \times r = (r^2)$, so that the measure of the interior of a circle is approximately $3r^2$.

The exact measure of the circular region is defined

$$m(\text{circular region}) = \pi r^2$$

where *r* is the measure of the radius of the circle. We may approximate this measure by using 3.14 as the value for π:

$$m(\text{circular region}) \approx 3.14\, r^2$$

Problem and Activity Set 5.4

1. Find (1) the perimeter and (2) the area of each circle for which the length of the radius is given. Use $\pi = 3.14$.
   - a. 6 inches
   - b. 3 feet
   - c. 15 yards
2. Find the missing data. Use $\pi = 3.1$.

CIRCLE	RADIUS	DIAMETER	CIRCUMFERENCE	AREA
A	4 IN.			
B		10 IN.		
C			200 FT.	

3. Find the area of the shaded portion of the figure below. Segment AB is a diameter of circle P and m($\overline{AB}$) = 2.

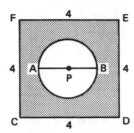

4. The figure below is a simple closed curve which is the union of a semi-circle and a diameter of this circle.

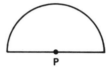

The measure of the area of the interior of this simple closed curve is 10. Do *not* use an approximation for π in order to answer the following.
   a. What is the circumference of circle P?
   b. What is the radius of circle P?
   c. What is the length of the semi-circle of circle P?
   d. What is the total length of the simple closed curve?
5. Which has the greater frying surface—a seven-inch frying pan or an eight-inch square skillet?
6. The moon is approximately 250,000 miles from the earth. The orbit of the moon around the earth represents approximately a circle. Suppose this orbit were a circle; then the path would lie in a plane and there would be an interior of the orbit (in the plane). Using π = 3.14, what would be an estimate for the area of this interior?
7. A wheel moves a distance of 15 feet along a track when the wheel turns once. What is the radius of the wheel?
8. The measure of the diameter of a given circle is 40. The circle is separated by points into eight arcs of equal measure. Use π = 3.14.
   a. What is the perimeter of the circle?
   b. What is the measure of each arc?
   c. What is the arc measure of each arc?
   d. On this circle, what is the linear measure of an arc of ten degrees?

9. Consider a circle with radius of measure r, and points $ABCD$ of the circle such that $m(\widehat{AB}) = m(\widehat{BC}) = m(\widehat{CD}) = m(\widehat{DA})$.

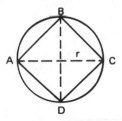

a. Using the $m(\widehat{AB})$ as the measure of the base of a triangle, determine the measure of the polygon.
b. Use eight points instead of four.
c. Use 16 points instead of eight.

10. a. Draw a circle congruent to the one illustrated on your paper. Cut out the figure of the circle. Then cut the representation of the circle into four pieces and reassemble these pieces as in figure b.

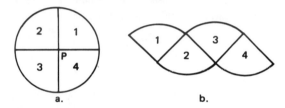

b. Next cut each of the four pieces into two congruent pieces. You now have eight congruent pieces of paper. Arrange these eight pieces of paper in a manner similar to that above. Your reassembled pieces should be similar to the following.

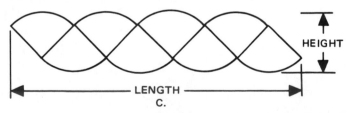

Using your ruler, find the height and length of the cutout indicated above. Then find the area represented by the pieces of paper. Determine the length and height to the nearest 1/8 inch.

c. Next, again cut each piece of paper into two congruent pieces and arrange as in figure c. You should now have 16 congruent pieces of paper.

   d. Find the height and length of this cutout and determine the area of the
      cutout as done in *b*.
   e. To answer the following questions, let $\pi = 3.14$.
      1. Find the circumference of circle *P*.
      2. Find the area of circle *P*.
      3. Is the length of the cutout obtained in *d* approximately 1/2 the
         circumference of the given circle?
      4. Is the height of the cutout obtained in *d* approximately that of the
         length of the radius?
      5. Subtract the measure of the area obtained in *d* from the measure of
         the area of circle *P*. What is the remainder?
      6. Do you think you could get a closer approximation of the area of
         circle *P* if 32 congruent pieces of paper were used instead of 16?

5.5 THE SPHERE

Recall that a circle is a set of points in a plane. The points of a *sphere* are
points of *space*. The reader should note the many "parallels" between a circle
and a sphere.

Definition 5.4—A Sphere

Given a point *P* and a segment $\overline{AB}$, the set of all points, *X*, such that
$\overline{PX} = \overline{AB}$ is a *sphere*.

Thus, we may think of a sphere as the set of points represented by the surface
of a basketball or a baseball. Only the part we can paint is the model of a sphere.

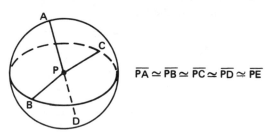

$$\overline{PA} \simeq \overline{PB} \simeq \overline{PC} \simeq \overline{PD} \simeq \overline{PE}$$

Figure 5.22

The point *P* is called the *center* of the sphere and the segments $\overline{PA}$, $\overline{PB}$, $\overline{PC}$,
etc. are called *radii* of the sphere. For any two radii of the sphere, $\overline{PC}$ and $\overline{PD}$, if
$\overline{PC} \cup \overline{PD} = \overleftrightarrow{CD}$, then $\overline{CD}$ is called a *diameter* of the sphere.

Recall that a circle separates the plane of the circle into the exterior and
interior of the circle. A sphere will separate *space* into the exterior and interior
of the sphere.

Again, we frequently denote the length of the radius of the sphere as r linear units. Thus, the sphere with center P and radius r linear units is the set of all points, X, $m(\overline{PX}) = r$. The exterior is the set of all points Y such that $m(\overline{PY}) > r$.

Similarly, if two or more distinct spheres have the same point P as center, they are *concentric spheres*.

Given a sphere with center point P and a plane α, exactly one of the following is true.

a. Sphere $P \cap$ plane $\alpha = \{\ \ \}$.
b. Sphere $P \cap$ plane α is exactly one *point*, or
c. Sphere $P \cap$ plane α is exactly one circle.

The case where the plane α intersects the sphere at exactly one point is illustrated in Figure 5.23. The point of intersection is point A.

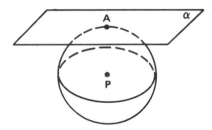

Figure 5.23

Definition 5.5—A Tangent Plane

Given a sphere with center P and a plane α, if sphere $P \cap$ plane α is exactly one point, the plane α is called a *tangent plane* to the sphere. If sphere $P \cap \alpha =$ point A, then A is the *point of tangency*.

There exist an infinite number of such planes that are tangent to a given sphere.

If we "lower" the plane α, as illustrated in Figure 5.24, the intersection of the plane and the sphere is a circle. Several of these circles are illustrated.

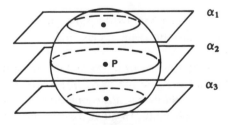

Figure 5.24

The plane α_2 contains the point P as an element. The circle which is the intersection of the sphere and the plane containing the center of the sphere is called a *great circle*.

Definition 5.6—A Great Circle of a Sphere

Given a sphere P and a plane α so that sphere $P \cap$ plane α is a circle,
if P is an element of α, then the circle is a *great circle* of sphere P.

There are an infinite number of these great circles of a sphere. All other circles determined by the plane passing through a sphere are called *small circles*.

Problem and Activity Set 5.5

1. If we think of the surface of the earth as a sphere, the line passing through the poles of the earth is called the axis of the earth. The diameter denoted by this axis intersects the sphere at the North and South Poles. The points represented by these poles are "directly opposite" each other. These two points are called *antipodal points*. Each diameter of a sphere will contain two such points.
    a. Is there an antipode of any given point of a sphere?
    b. Is there more than one antipode of any given point of a sphere?
    c. How many great circles may contain a given point of a sphere?
    d. How many small circles contain a given point of a sphere?
    e. Can a small circle contain a pair of antipodal points of a sphere?
    f. On a sphere, must the intersection of every two great circles contain exactly two points?
    g. On a sphere, must the intersection of any two small circles be exactly two points?
2. Let P_1 and P_2 be distinct points of a sphere which are not antipodes. How many great circles contain these two points?
3. When we think of the surface of the earth as a sphere, the equator represents a great circle. Small circles whose planes are parallel to the plane of the equator are called *parallels of latitude*. These small circles have centers on the axis of the earth and their planes are perpendicular to this axis. If $m(\angle EPA)$ = 30, we say the circle with center O is the circle whose latitude is $30°$ north of the equator. Any point on this circle is said to have $30°$ north latitude $(30°N)$.

    Great circles which contain the North Pole and the South Pole are called *circles of longitude* and the semi-circles such as $\overparen{NAS}$, $\overparen{NBS}$, and $\overparen{NCS}$ are called *meridians*.

    In order to name meridians, we pick one and designate it the *prime meridian*. Let us pick $\overparen{NBS}$. In the figure on page 147, if $m(\angle BPA) = 40$, we name the meridian $\overparen{NAS}$ as the meridian whose longitude is $40°$ West. Any point on $\overparen{NBS}$ has $40°$ West Longitude. In all maps, the prime meridian is the one which passes through Greenwich, England.

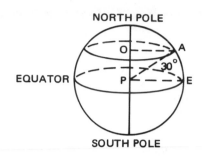

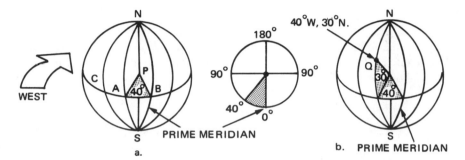

The point denoted on the sphere as Q has longitude $40°$ West and latitude $30°$ North. ($40°$W, $30°$N.).

Using a globe, find the approximate location of each of the following cities. Indicate their position by listing the longitude first, followed by their latitude.

a. San Francisco   b. New York
c. Dallas   d. Moscow
e. Peking   f. Paris
g. The coordinates denoting your hometown.

4. Using a map or globe, find the cities located by the following coordinates:
    a. $74°$ W., $40°$ N.   b. $90°$ W., $30°$ N.
    c. $145°$ E., $37°$ S.   d. $57°$ W., $13°$ S.

5. Determine the coordinates of the antipodal point of a given point denoted by $90°$W., $45°$N., *without* using a globe.

6. Find the coordinates of the antipodal points of each of the following.
    a. $80°$ W., $40°$ S.   b. $180°$ W., $45°$ N.
    c. $30°$ E., $50°$ N.   d. $160°$ E., $60°$ S.

7. The intersection of the Greenwich meridian and the equator is labeled $0°$. From this point, we may "follow" the path of the equator east, or west, until we reach the meridian which contains the antipode of this point. This antipodal point is marked $180°$.

   As you know, 24 hours is the time it takes the earth to make one revolution about its axis. In order to determine this, we "mark" a point on the surface of the earth with reference to the sun. When the "marked point" again returns to the same reference position, the time *one day* is defined. By this procedure, the earth "turns" 360° every 24 hours. This is a turn of 15° per hour.

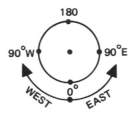

When the Greenwich meridian is precisely "in line" with the sun's rays, the time is defined to be noon *Greenwich Mean Time* (GMT). This time is denoted 12:00. *Other times are denoted using a 24-hour clock*, not using a 12-hour clock. Thus, 1:00 P.M. on a 12-hour clock is 13:00 on a 24-hour clock; 6:00 A.M. is 06:00; 12 midnight is denoted 24:00 or 0:00; 10:00 P.M. is denoted 22:00.

TIME ZONES. Suppose it is presently 12:00 at every point of the 0° (Greenwich) meridian. What time will it be at every point of the 15° E. meridian? Since the earth rotates on its N-S axis 15° every hour, the points of the 15° E. meridian were directly "in line" with the sun's rays exactly one hour ago. We say that the time at every point X of the 15° E. meridian is is 13:00 (1:00 P.M.).

Similarly every point of the 15° W. meridian will be "in line" with the sun's rays after a lapse of one hour. Therefore, the time at every point of the 15° W. meridian is 11:00 (11:00 A.M.).

Hence, we see a need for establishing "time zones," in order that the points of any meridian be "in line" with the sun's rays at 12:00 (noon). These time zones are generally bounded by the meridians 0°, 15° E, 30° E, 45° E, etc., and by the meridians 0°, 15° W., 30° W., 45° W., etc.

The continental United States is partitioned into four different time zones. These zones are the Eastern, Central, Mountain, and Pacific Zones. These are illustrated below.

If it is 12:00 in New York, it is 11:00 in Dallas, 10:00 in Denver, and 9:00 in San Francisco.

THE INTERNATIONAL DATE LINE. Assume two students, A and B, live in a town with coordinate $0°$; the time is 12:00, June 1, 1973. Furthermore, assume that student A walks in a westerly direction and student B walks in an easterly direction from $0°$, each student following the path of the equator until they meet at the antipodal point, $180°$, twenty-four hours later.

When student A "arrives" at the meridian $15°$ W., exactly two hours will have elapsed. Since this point will be exactly "in line" with the sun's rays, the student must re-set his watch to read 12:00. When he "arrives" at the $30°$ W. meridian, he must again reset his watch to read 12:00. Continuing in this manner, he will arrive at the $180°$ meridian *exactly the same hour of the same day that he left the $0°$ meridian* (12:00, June 1, 1973).

But what has happened to student B and his easterly trek? When student B arrived at the $15°$ E. meridian, the time was 14:00; when he arrived at the $30°$ E. meridian, the time was 16:00.

When student B arrived at the $90°$ E. meridian, the time was 24:00. *Student B must now flip the leaf of his calendar to read June 2, 1973.* Continuing in this manner, student B will arrive at the $180°$ meridian (and meet student A) at *12:00, June 2, 1973.*

Hence, we see a need for establishing an *International Date Line* which approximately follows the $180°$ meridian. If a traveler crosses this imaginary line when progressing in an easterly direction, he must *subtract* a day when crossing the date line; if traveling in a westerly direction, he must *add* a day when crossing the date line.

Using a globe or map of the world, answer the following questions.
a. When it is 12 noon in New York, what time is it in Los Angeles? in Moscow?
b. What time is it in New York when it is 12 noon in Moscow?
c. Fred's home is 1,400 miles due west of New York. How many hours earlier (or later) does the sun rise at Fred's home than in New York?
d. If the time is 11:00 A.M. at the location denoted by the coordinates $40°$W., $40°$N., what time is it at the antipodal point?
e. Assume it is 12:00 noon, July 4, 1973. What is the time and the date in Paris if one's location is $40°$ W., $40°$ N.?

5.6 VOLUME AND SURFACE AREA OF A SPHERE

Recall that a cube has a "surface," and when we speak of the volume of a cube we mean the volume of a rectangular solid whose surface is a cube. Similarly, by volume of a sphere, we mean the volume of a solid whose surface is a sphere. We determine the measure of the volume of a sphere using the cube as the unit of measure. Hence, we are placed in a position that parallels the situation for determining the circumference of a circle. It is very difficult, and beyond the scope of this text, to derive a formula to determine the volume of a sphere. Mathematicians have proved that the volume of a sphere may be found by using the formula

$$V = 4/3\pi r^3$$

where r is the measure of the radius of the sphere.

Similarly, it is very difficult to determine the formula for the area of the surface of a sphere. Again, mathematicians have proved that this formula is

$$A = 4\pi r^2$$

where r is the measure of the radius of the sphere.

Problem and Activity Set 5.6

1. For each sphere whose radius is given, find the volume of the corresponding spherical solid. Use 3.14 as the approximate value of π.
   a. $r = 20$ inches   b. $r = 3$ feet
   c. $r = 5$ yards   d. $r = 16$ centimeters
2. For each sphere of Problem 1, find the surface area.
3. An oil tank is in the shape of a sphere whose diameter is 100 feet.
   a. If paint costs $10 per gallon and one gallon covers 300 square feet, what is the cost of paint for the surface?
   b. If oil costs $.33 per gallon, find the value of the oil in the tank when the tank is half full (1 cubic foot = 7 1/2 gallons).
4. When the radius of a sphere is doubled, what effect does this have on the surface area? on the volume?
5. Two spheres have diameters of ratio 3/2.
   a. Find the ratio of their volumes.
   b. Find the ratio of their surface areas.

6

PLANE COORDINATE GEOMETRY

6.1 INTRODUCTION

Frequently, we must "take trips in a plane." We may need to take a trip from home to school, to a movie, to a football game, or to some activity which occurs "across town."

As you know, we can locate any building in a city by giving its distances from two streets that are perpendicular. Let Figure 6.1 represent a map of Midwest City. Since Midwest City is also known as a vacation resort, assume that a policeman is standing in the intersection of Central Avenue and Main Street, giving information to visitors regarding places of interest.

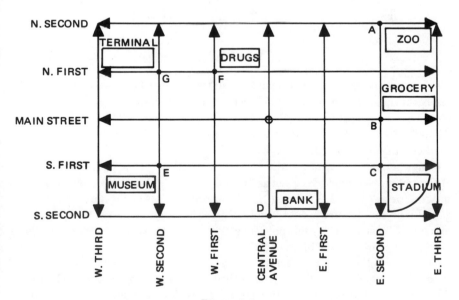

Figure 6.1

The policeman may inform the visitors, "To get to the zoo, travel two blocks east on Main Street and then two blocks north on East Second Avenue." To tell visitors how to get to the museum, he may state, "Travel two blocks west on Main Street and then one block south on West Second Street."

It may be easier for the policeman to provide instructions if we think of the streets of Midwest City as lines. If we do this, we then have a "grid" for the policeman to use as reference.

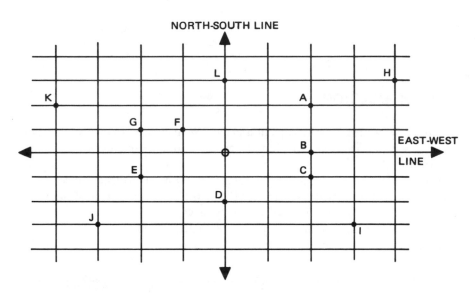

Figure 6.2

Let us assume that the policeman first states the east-west direction and then states the north-south direction. We will denote an easterly direction of four blocks by the special symbol "$\overrightarrow{4}$" and a westerly direction of two blocks by "$\overleftarrow{2}$." Similarly, we denote a northerly direction of three blocks by "3↑" and a southerly direction of four blocks by "4↓."

Using these special symbols, the various instructions of the policeman located at Central and Main may be given as "East 2, North 2, $(\overrightarrow{2}, 2↑)$" in order to get to the zoo, or as "West 2, South 1, $(\overleftarrow{2}, 1↓)$" in order to get to the museum. On the grid, these are denoted as "*A*" and "*E*" respectively. In order to get to the library, "*L*" on the grid, the policeman would state "(0, 3↑)," meaning "No blocks east or west, three blocks north."

Thus, the policeman, in order to describe a trip on this "grid" must always state two numbers. These numbers are called *ordered pairs of numbers*. The term "ordered pair" is a very descriptive term. To say "ordered pair" is to say that it makes a difference in which order the numbers are considered.

The location of any non-empty intersection of these lines may be designated by one particular ordered pair. The location "*D*" would be designated (0, 2↓), which means "zero blocks east, 2 blocks south." "*F*" would be (1̄, 1↑); "*B*" would be (2⃗, 0).

Problem and Activity Set 6.1

1. Using the accompanying diagram list the ordered pair of numbers which describes the "trips" indicated. Recall that the first member of the ordered pair denotes an east-west direction; the second member denotes a north-south direction.

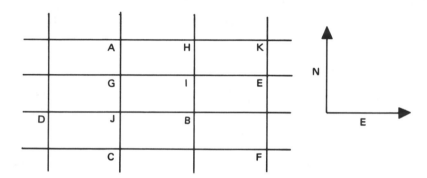

   a. From *D* to *H*   b. From *B* to *F*
   c. From *G* to *C*   d. From *D* to *B*
   e. From *B* to *H*   f. From *F* to *A*

2. Using the grid of Problem 1, start at the given point and take the "trip" which is denoted by the ordered pair. Name the point at which each trip terminates.
   a. Start at *I*, take trip (1⃗, 2↓)
   b. Start at *B*, take trip (2⃖, 0)
   c. Start at *C*, take trip (1⃗, 3↑)
   d. Start at *A*, take trip (0, 3↓)
   e. Start at *J*, take trip (2⃗, 1↓)

6.2 THE GRAPH OF POINTS

On graph paper (which we may think of as a plane), we can locate any point in this plane by using a technique similar to that used by the policeman in the previous section. If we think of Main Street and Central Avenue as perpendicular *number lines* with the policeman standing at the intersection "(0, 0)" of these lines, we can now represent the map of Midwest City in the following manner.

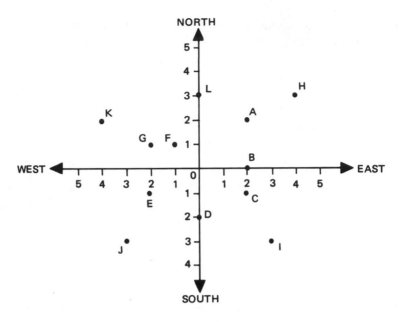

Figure 6.3

On this map, we would locate the point *A* by the ordered pair $(\overrightarrow{2}, 2\uparrow)$, point *K* by $(\overleftarrow{4}, 2\uparrow)$, point *D* by $(0, 2\downarrow)$. The intersection of Main Street and Central Avenue is called the *origin* and is designated by the ordered pair $(0, 0)$. If we think of Main Street as a horizontal number line, we could, by starting at zero, designate points on this line by using the integers. The positive integers would represent an easterly direction and the negative integers would represent a westerly direction, as illustrated in Figure 6.4.

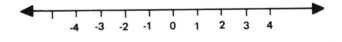

Figure 6.4

Using this number line, a trip of three blocks west would be designated "−3," a trip of four blocks east would be designated "4." If we follow the same scheme and let Central Avenue be represented by a vertical number line, then a trip of two blocks north would be designated "2," a trip of three blocks south would be referred to as "−3."

If we represent horizontal and vertical number lines whose intersection is exactly one point, the two lines lie in the same plane. If one line is horizontal and the other is vertical, they are perpendicular. Recall that their point of intersection is called the *origin*.

Figure 6.5

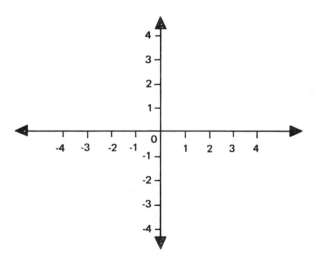

Figure 6.6

We can designate points in this plane by the use of ordered pairs. The first member of the ordered pair will indicate a horizontal movement, and the second member will indicate a vertical movement. By convention, we agree that if the first member is positive, the movement will be to the right; if negative, the movement will be to the left. Also, if the second member is positive, the movement will be up; if negative, the movement will be down. Therefore, the ordered pair (2,3) would designate a point which is two units to the right of the origin and three units up. To locate this point, you would start at the origin, move two units to the right, and then three units up. We might refer to this point as $P(2,3)$, which means the point designated by the ordered pair (2,3).

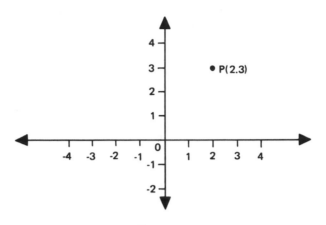

Figure 6.7

The first member of the ordered pair of numbers is called the *abscissa* of any point P. We may think of this number as denoting the number of units to the "left or right" from the vertical line. This first member is generally symbolized by the letter "x."

The second member of the ordered pair is called the *ordinate* of point P. We may think of this number as denoting the number of units "up or down" from the horizontal line. This second number is generally symbolized by the letter "y."

The members of the ordered pair are called the *coordinates* of P. Several points are designated along with their corresponding coordinates in Figure 6.8.

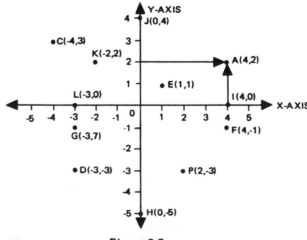

Figure 6.8

Since "x" designates "right or left," we call the horizontal number line the x-axis, and since "y" designates "up or down," we usually call the vertical line the y-axis.

In Figure 6.8, the abscissa of A is 4, the ordinate of A is 2, and the coordinates of A are (4,2). This is denoted "A(4,2)."

The coordinates of any point are called *Cartesian coordinates* after the French mathematician René Descartes. Among other things, his works provided a foundation for relating the separate topics of algebra and geometry.

Problem and Activity Set 6.2

1. Obtain a piece of graph paper. Designate a horizontal number line as the x-axis and a vertical number line as the y-axis. Locate each point designated by the ordered pair.
   a. $A(2,4)$
   b. $B(-2,5)$
   c. $C(2,-5)$
   d. $D(-4,-4)$
   e. $E(0,-2)$
   f. $F(2,0)$
   g. $G(0,0)$
   h. $H(-2,0)$
   i. $I(0,3)$

2. Determine the coordinates of each point named below.

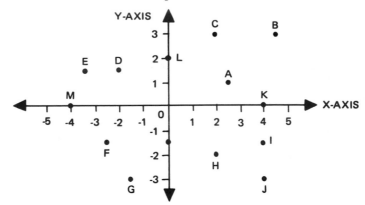

3. With perpendiculars drawn as shown below, what are the coordinates of A, B, and C?

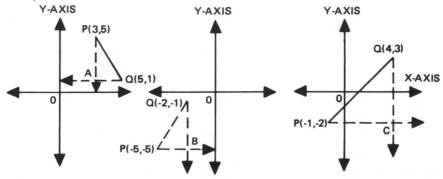

4. On your graph paper, draw the triangles ABC whose vertices are as follows:
   a. $A(3,-2)$, $B(-2,-5)$, $C(0,1)$
   b. $A(0,0)$, $B(-6,6)$, $C(6,4)$
   c. $A(4,-2)$, $B(4,-5)$, $C(0,4)$
5. On your graph paper, draw the quadrilaterials $ABCD$ whose vertices are as follows:
   a. $A(-3,0)$, $B(2,4)$, $C(6,0)$, $D(3,5)$
   b. $A(0,0)$, $B(3,5)$, $C(8,0)$, $D(5,7)$

6.3 INCLINATION

Recall that a line in a plane is a particular subset of points of the plane, and that any two points of this plane determine a unique line. In Figure 6.9, consider the two points $(3,4)$ and $(-2,0)$, and the line ℓ_1 determined by these two points. Other lines illustrated are lines ℓ_2 containing $(-4,2)$ and $(4,0)$ and ℓ_3 containing $(0,-5)$ and $(8,-5)$.

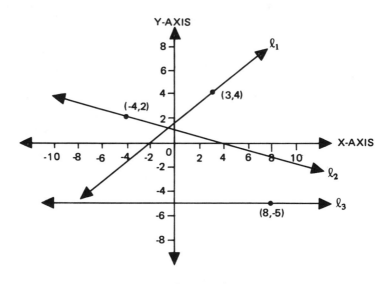

Figure 6.9

We may intuitively think of the "steepness" of a line ℓ by considering the x-axis as a horizontal line. When this is done, reference is made to the *inclination* of the line.

The measure of the angle determined by the ray containing the positive portion of the x-axis and the ray of the line that is "above" the x-axis is called the inclination of the line. The reader should verify that the inclination of ℓ_1 of Figure 6.9 is approximately 42, and the inclination of ℓ_2 is approximately 161, where the unit is degrees.

If we say the measure of an angle is 40, we will interpret this to mean that a degree is the unit of measure. Therefore, for clarity's sake, we will speak of the inclination as being $40°$ and understand that we name the unit to avoid ambiguity.

Recall that for any two given lines of the same plane, ℓ_1 and ℓ_2, $\ell_1 \cap \ell_2 = \{ \quad \}$ or $\ell_1 \cap \ell_2$ is exactly one point. In Figure 6.9, $\{x\text{-axis}\} \cap \ell_3 = \{ \quad \}$. The inclination of ℓ_3 is 0. In general, for any line ℓ, if $\ell \cap \{x\text{-axis}\} = \{ \quad \}$, then the inclination of ℓ is 0.

But $\ell_1 \cap \{x\text{-axis}\}$ is exactly one point and $\ell_2 \cap \{x\text{-axis}\}$ is exactly one point. In order to determine the inclination of these lines, consider the case of any line ℓ of the plane, $\ell \cap \{x\text{-axis}\} = P$.

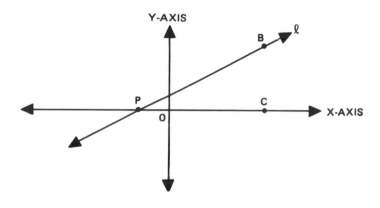

Figure 6.10

Let B be a point of ℓ so that B is "above" the x-axis and let C be a point of the positive x-axis. Then the inclination of ℓ is determined by $\angle CPB$. The reader may verify that the inclination of line ℓ is approximately 22.

6.4 SLOPE OF A LINE

If we were to stand at the base of a mountain and look toward the peak, we might remark, "That mountain is very steep," or "That mountain is not very steep." We could say similar things about the roof of a house. When we say the word "steep," we want to convey the idea of how quickly something "rises" above the horizontal. If we were to think of a line in a plane, the steepness of this line would be judged on how quickly it rises above a horizontal line. Mathematicians have agreed to refer to the steepness of a line in a plane by using the word *slope*.

To determine the measure of the slope of a line, we would begin at a point of a line, move horizontally to the right one unit, and observe the point of the line directly above, or below, that point. For example, if we move one unit to the right and observe that the line is one unit above, we would say the line has slope 1. If the line were two units below this point, we would say the line has slope −2.

Then slope can be thought of as a number which describes a relationship between horizontal and vertical, that is, it is a *ratio*. This is most often thought of as the ratio of the "rise to the run." This is expressed as the ratio of the vertical change to the horizontal change.

Definition 6.1—The Slope of a Line.

The *slope* of a line ℓ determined by the distinct points $P_1 (x_1, y_1)$ and $P_2(x_2, y_2)$ is the number:

$$m = \frac{y_2 - y_1}{x_2 - x_1} = \frac{y_1 - y_2}{x_1 - x_2} = \frac{\text{vertical change}}{\text{horizontal change}}$$

In Figure 6.11b, the slope of line ℓ is

$$m = \frac{3 - 1}{3 - 1} = \frac{2}{2} = 1$$

or

$$m = \frac{1 - 3}{1 - 3} = \frac{-2}{-2} = 1$$

We think of a line as "sloping upward" from left to right, or as a line "sloping downward" from left to right. The reader should note that the absolute value of the number m indicates the "steepness" of the line.

The slope of line ℓ in Figure 6.12a is:

$$m = \frac{1 - 0}{4 - 0} = \frac{1}{4}$$

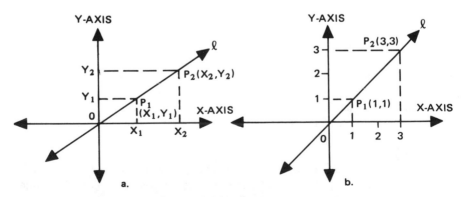

Figure 6.11

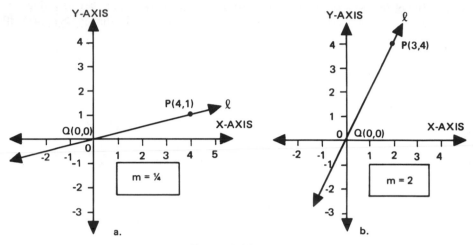

Figure 6.12

The slope of line ℓ in Figure 6.12b is:

$$m = \frac{4-0}{2-0} = 2$$

If the line "slopes upward" from left to right, the slope of the line is a positive number because the numerator and denominator are both positive numbers, i.e., in the ratio

$$m = \frac{y_2 - y_1}{x_2 - x_1}$$

$y_2 > y_1$ and $x_2 > x_1$ (or both numerator and denominator are negative if we reverse the order of the coordinates). This is illustrated in Figure 6.13, where $y_2 = 2, y_1 = -1, x_2 = 1$, and $x_1 = -3$. The reader should note that $2 > -1$ and $1 > -3$.

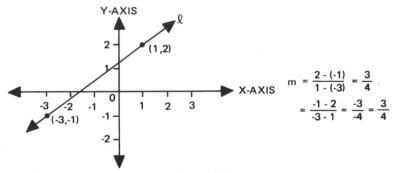

Figure 6.13

If the line ℓ "slopes downward" from left to right, the number denoting the slope is negative. This is so because for $x_1 < x_2$, we find that $y_1 > y_2$. (Or if $x_1 > x_2$, then $y_1 < y_2$ if we reverse the order of the coordinates).

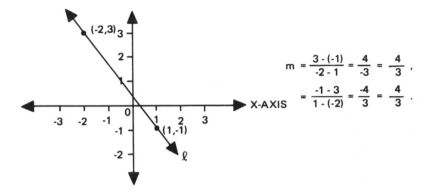

$$m = \frac{3-(-1)}{-2-1} = \frac{4}{-3} = \frac{4}{3} \, ,$$

$$= \frac{-1-3}{1-(-2)} = \frac{-4}{3} = \frac{4}{3} \, .$$

Figure 6.14

A line ℓ may also have slope 0, this occurs if $y_2 = y_1$, thereby making the numerator of the ratio zero. This is illustrated in Figure 6.15.

All lines with slope 0 are called horizontal lines. The reader should note that the slope of the x-axis is 0, and is therefore a horizontal line.

A ratio can be thought of as a comparison of two numbers by division. Therefore, if this interpretation is accepted, the ratio "4 to 0" is undefined, since division by zero is undefined.

Thus, if we consider the line determined by $P_1(x_1, y_1)$ and $P_2(x_2, y_2)$, if $x_1 = x_2$, the slope of this line would be undefined since $x_2 - x_1 = x_1 - x_2 = 0$. The slope of any vertical line is undefined.

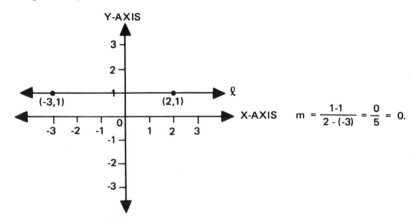

$$m = \frac{1-1}{2-(-3)} = \frac{0}{5} = 0.$$

Figure 6.15

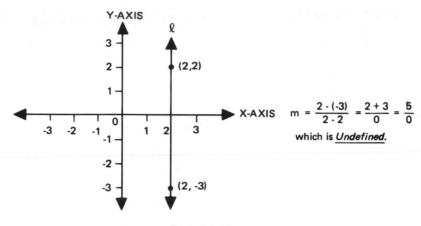

$$m = \frac{2 - (-3)}{2 - 2} = \frac{2 + 3}{0} = \frac{5}{0}$$

which is *Underfined.*

Figure 6.16

Problem and Activity Set 6.3

1. Using your protractor, find the inclination of each line to the nearest degree.

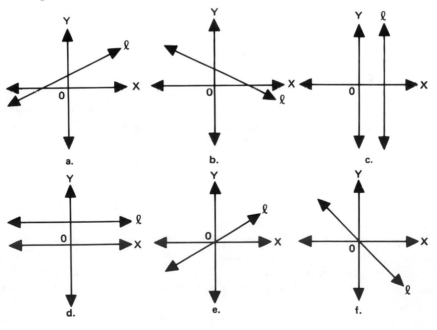

2. Replace the "?" in such a way that the line containing the two points will be horizontal.
   a. (3,2) and (-7,?)

  b. (-4,1) and (8,?)
  c. (0,0) and (?,4)
3. Replace the "?" in such a way that the line containing these points will be vertical.
  a. (?,3) and (6,-4)
  b. (4,-4) and (?,5)
  c. (0,0) and (?,7)
4. A road "rises" upward three feet for every 60 feet of horizontal distance. What is its slope?
5. Determine the slope of the segment containing each of the following points.
  a. (0,0) and 3,5)   b. (3,4) and (6,11)
  c. (0,0) and (-4,-2)   d. (0,0) and (-4,-3)
  e. (-1,-2) and (-5,-7)   f. (-5,7) and (8,-3)
6. Replace the "?" in such a way that the line containing the two points will have the given slope. (Hint: Substitute in the slope formula.)
  a. (0,0) and (?,4); $m = 4$.
  b. (2,2) and (6,?); $m = -2$.
7. Is the point $C(1,2)$ an element of the line containing $A(-2,-2)$ and $B(3,4)$? (Hint: Is the slope of $\overline{CB}$ the same as that of $\overline{AB}$?)
8. Is the point $(2,-1)$ an element of the segment containing $(-5,4)$ and $(6,-8)$?
9. Determine the slope of a segment containing the points
  a. $(0,n)$ and $(n,0)$, where n is an integer
  b. $(2d,-2d)$ and $(0,d)$, where d is an integer
  c. $(a + b, a)$ and $(a - b, b)$, where a, b are integers
10. On your graph paper, draw the x-axis and the y-axis and plot the points $A(-3,2)$, $B(-2,-2)$, $C(4,-2)$, and $D(3,2)$. Draw the quadrilateral $ABCD$. Find the slope of each side of $ABCD$. Which sides have the same slope?

6.5 EQUATIONS

The reader will recall that an expression of the type $4 + 3 = 7$ is called an equation. Likewise, an expression of the type $x + 3 = 7$ is called an equation. Most often in equations of the type $x + 3 = 7$, we are interested in the truth set of this equation. By truth set, we mean the set of all numbers such that if x is replaced by each member of this set then $x + 3 = 7$ is a true sentence. The equation $x + 3 = 7$ is an equation with a single "variable." The variable in this instance is x.

Let us now consider an equation of the type $x + y = 10$. Here we see that two variables are included. Now let us consider the truth set of this equation. If we replace x with 3 and y with 7 then the sentence is true. Then the truth set consists of number pairs (more precisely called ordered pairs). Then other pairs are seen to be in the truth set, e.g. (6,4), (8,2) and (4,6). In fact, depending on the domain, other pairs might be (-3, 13), (24, -14), and (6 ⅓, 3 ⅔).

Problem and Activity Set 6.4

1. Replace the "□" in such a manner that the ordered pairs are members of the truth sets of the given equations. Let the set of Integers be the domain.

A) x + 2y = 10

X	Y
0	□
1	□
□	3
□	0
-1	□
□	-2

B) 2x + y = 20

X	Y
□	0
5	□
□	-2
□	-10
-7	□
8	□

2. Given the following members of the truth set, find the equation.

A) $\Delta x + \Box y = 4$

X	Y	
8	4	4
10	6	4
0	-4	4
12	8	4
-6	-10	4
-10	-14	4

B) $\Delta x + \Box y = 8$

X	Y	
0	8	8
1	6	8
-1	10	8
4	0	8
5	-2	8
-7	22	8

C) $\Delta x + \Box y = 6$

X	Y	
1	-3	6
0	-6	6
-2	12	6
2	0	6
3	3	6
-3	-15	6

6.6 DESCRIBING A LINE BY AN EQUATION

Consider a non-vertical line ℓ in a plane with slope m. Let $P(x_1, y_1)$ be a point of ℓ. Let $Q(x, y)$ be a point of ℓ different from $P(x_1, y_1)$. Then the slope of the line ℓ is determined by the ratio:

$$m = \frac{y - y_1}{x - x_1}$$

Since ℓ is non-vertical, we know $x_1 \neq x$. If we multiply both members of this equation by $x_1 - x$, we get the equation

$$y - y_1 = m(x - x_1) \qquad \qquad (Eq.\ 1)$$

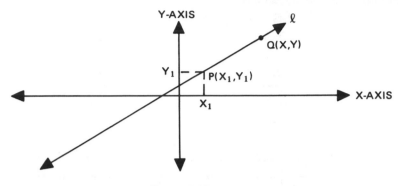

Figure 6.17

Suppose we know that a line contains the point (3, 5) and has a slope of 2. Of interest are the other points of this line. So letting $x_1 = 3$ and $y_1 = 5$ and $m = 2$, Eq. 1 then becomes:

$$y - 5 = 2(x - 3)$$

If we solve this in the usual manner, we find:

$$y = 2x - 1$$

Some pairs of the truth set of this equation are (1,1), (2,3), and (4,7). These are points of the line that contain (3,5) and has a slope of 2. The equation $y = 2x - 1$ is called the *equation of the line*. Notice that the coefficient of the x is the slope.

If you were to use any two points of the truth set of this equation and look at the line determined by these two points, you would notice that the line intercepts the y-axis at the point $(0, -1)$. The number -1 is called the y-intercept. This is illustrated in Figure 6.18.

Problem and Activity Set 6.5

1. In each of the following, the coordinates of a point P and the value of slope m is given. Find the equation of the line.
   a. $P = (1,1); m = 4$
   b. $P = (-1,1); m = -2$
   c. $P = (0,5); m = 5$
   d. $P = (-1,-3); m = \frac{1}{2}$
2. Find five members of the truth set for each equation of a.–d. above.
3. Find the equation for each of the following. Determine (1) the slope of each equation and (2) the y-intercept of each equation.

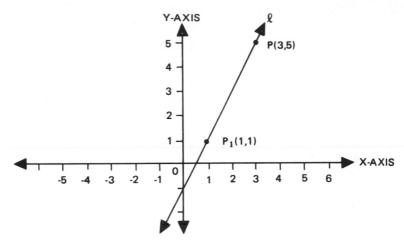

Figure 6.18

a. The line containing $A(0,0)$ and $B(2,2)$
b. The line containing $A(-1,2)$ and $B(2,5)$
c. The line containing $A(-1,-1)$ and $B(6,-1)$
d. The line containing $A(1,0)$ and $B(1,7)$

4. Find the equation of each line if the slope and y-intercept of each line is given below.
   a. $m = 2$, y-intercept = 4
   b. $m = \frac{1}{2}$, y-intercept = 0
   c. $m = -3$, y-intercept = 3
   d. $m = 0$, y-intercept = 4

5. The vertices of triangle ABC are $A(1,1)$, $B(0,0)$, and $C(-2,2)$. Find the equation of each side of $\triangle ABC$.

6.7 DESCRIBING A LINE BY A GRAPH

If we let m be the slope of a line, the general form (called the point-slope form) of the equation of the line is $y = mx + b$. You probably will have noted that the number represented by b tells you where the line intercepts the y-axis. Appropriately the number b is called the "y-intercept."

Now let us consider the graphic representation of the truth set of the equation $y = mx + b$. This representation is called the "graph" of the equation. It is sufficient to determine two members of the truth set and then use these points in drawing the line. Remember, if you know the y-intercept, you already have one of these points. Since the graph of the equation $y = mx + b$ is a line, this equation is called a linear equation.

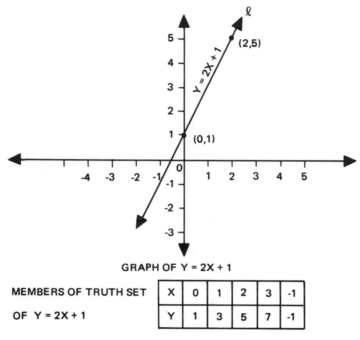

GRAPH OF Y = 2X + 1

MEMBERS OF TRUTH SET	X	0	1	2	3	-1
OF Y = 2X + 1	Y	1	3	5	7	-1

Figure 6.19

Problem and Activity Set 6.6

1. By changing to a point-slope form where necessary, show that the graph of each equation is a line. Find three points that are members of this line. Find the y-intercept of the line.

   a. $y = 3x + 4$ b. $2x + y = 7$
   c. $y - 2 = 2(x - 4)$ d. $x - 3y = 12$
   e. $y = x$ f. $y = 2x$
   g. $y = 2x - 4$ h. $y = 2x + 1$
   i. $x = 4$ j. $x = 0$
   k. $y = 0$

2. Find the equation for each of the following; graph the equation; find three elements of the truth set of each equation. Find the y-intercept of each line.

   a. The line containing $(1,1)$ with slope 3
   b. The line containing $(1,0)$ and $(0,1)$
   c. The line with slope 2 and y-intercept 4
   d. The x-axis
   e. The y-axis
   f. The horizontal line containing $(4,3)$
   g. The vertical line containing $(2,2)$

6.8 PAIRS OF LINES

If we investigate the truth sets of two (or more) linear equations, either the intersection is the empty set or the intersection contains exactly one ordered pair.

Consider the following under the stated conditions.

A. The Intersection of the Truth Sets is the Empty Set

Consider the truth sets of $y = 3x + 2$ and $y = 3x - 2$. Some ordered pairs of the truth sets, I_1 and I_2 are listed below. Also, the graph of each equation is illustrated.

$$I_1 = \left\{(x,y) \,|\, y = 3x + 2\right\} = \left\{\ldots, (-2,-4), (-1,-1), (0,2), (1,5), (2,8), \ldots\right\}$$

$$I_2 = \left\{(x,y) \,|\, y = 3x - 2\right\} = \left\{\ldots, (-2,-8), (-1,-5), (0,-2)\ (1,1), (2,4), \ldots\right\}$$

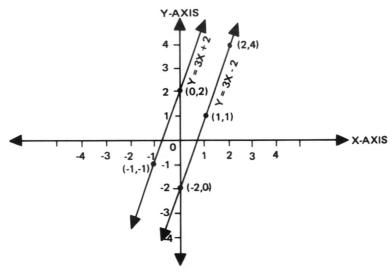

Figure 6.20

If we continue listing the elements within the truth sets, we will find no common pairs. If this is the case, the graphs of these linear equations are *parallel lines*.

Definition 6.2—Parallel Lines

Let ℓ_1 and ℓ_2 be coplanar lines with truth sets I_1 and I_2, respectively. If $I_1 \cap I_2 = \{\ \}$, then ℓ_1 and ℓ_2 are *parallel*.

You probably noticed that the slope of $y = 3x + 2$ and $y = 3x - 2$ is 3. Thus, we may say that if two lines ℓ_1 and ℓ_2 have the same slope, they are parallel; conversely, if two lines are parallel, they have the same slope.

EXAMPLE

Show that the line ℓ_1 containing $(2,0)$ and $(-6,4)$ is parallel to the line ℓ_2 containing $(2,-3)$ and $(-4,0)$.

Let m_1 be the slope of ℓ_1 and m_2 be the slope of ℓ_2.
Then:

$$m_1 = \frac{4 - 0}{-6 - 2} = \frac{4}{-8} = -\frac{1}{2}$$

$$m_2 = \frac{-3 - 0}{2 - (-4)} = \frac{-3}{6} = -\frac{1}{2}$$

Since $m_1 = m_2$, $\ell_1 \parallel \ell_2$.

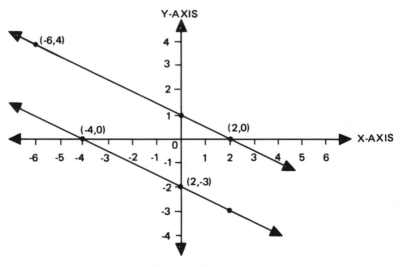

Figure 6.21

B. The Intersection of the Truth Sets is Exactly One Ordered Pair

Consider the truth sets, I_1 and I_2 of $y = x + 1$ and $y = -2x + 4$.

$$I_1 = \left\{(x, y) \mid y = x + 1\right\} = \left\{\ldots, (-2,-1), (-1,0), (0,1), \boxed{(1,2),} (2,3), \ldots\right\}$$

$$I_2 = \left\{(x, y) \mid y = -2x + 4\right\} = \left\{\ldots, (-2,8), (-1,6), (0,4), \boxed{(1,2),} (2,0), \ldots\right\}$$

$I_1 \cap I_2 = \{(1,2)\}$. The graph of these equations is illustrated in Figure 6.22. We say that the "point of intersection" is the point denoted by the coordinates (1,2).

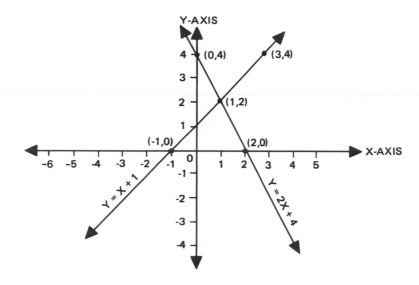

Figure 6.22

Of particular interest are two lines whose intersection is one point and which are perpendicular.

Definition 6.3—Perpendicular Lines

Let each of ℓ_1 and ℓ_2 be a line with truth sets I_1 and I_2 and slopes m_1 and m_2, respectively. If $I_1 \cap I_2$ is exactly one ordered pair and if $m_1 \times m_2 = -1$, then ℓ_1 and ℓ_2 are *perpendicular*.

In order to illustrate this, consider $y = 2x + 3$ and $y = -\frac{1}{2}x + 3$. Then the slope m_1 of $y = 2x + 3$ is 2 and the slope m_2 of $y = -\frac{1}{2}x + 3$ is $-\frac{1}{2}$.

$$m_1 \times m_2 = 2 \times (-\tfrac{1}{2}) = -1$$

The graphs of these equations are in Figure 6.23. The student may verify that the measure in degrees of $\angle 1 = \angle 2 = \angle 3 = \angle 4 = 90$, using the usual measurement techniques.

Hence, if two lines are perpendicular, the product of their slopes is -1, that is $m_1 \times m_2 = -1$. Conversely, if the product of the slopes of two lines is -1, the lines are perpendicular.

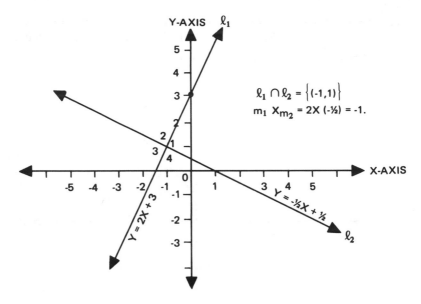

Figure 6.23

EXAMPLE

Show that the line ℓ_1 containing the points $(0,2)$ and $(-3,1)$ and the line ℓ_2 containing $(0,0)$ and $(-1,3)$ are perpendicular.

$$m_1 = \frac{2-1}{0-(-3)} = \frac{1}{3}$$

$$m_2 = \frac{3-0}{-1-0} = -3$$

$$m_1 \times m_2 = 1/3 \times -3 = -1$$

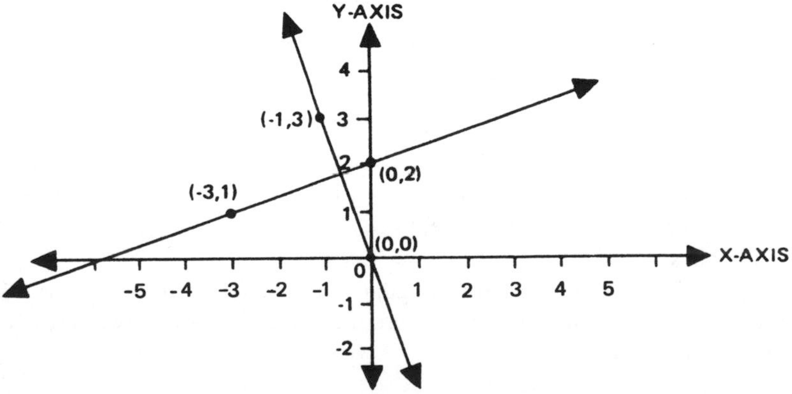

Figure 6.24

Problem and Activity Set 6.7

1. Find the common member of the truth sets of the following pairs of equations. Draw their graphs to verify your choice.
   a. $y = 2x$ and $x + y = 6$
   b. $y = 2x$ and $y - 2x = 3$
   c. $x + y = 4$ and $2y = 6 - 2x$
2. The graphs of which pairs of the equations listed below would be parallel lines? Verify your choices by drawing the graphs of these equations.
   a. $y = 3x + 1$ b. $y = 4x + 1$
   c. $2y = 8x + 2$ d. $y - 3x = 3$
3. The graphs of which pairs of the equations listed below would be perpendicular lines? Verify your choices by drawing the graphs of these equations.
   a. $y = 2x + 3$ b. $3y + x = 15$
   c. $12 = 2y + x$ d. $y = 3x + 2$
4. Show that triangle ABC is a right triangle with right angle at C if the vertices of $\triangle ABC$ are $A(3,0)$, $B(7,-2)$ and $C(4,1)$.
5. Show that quadrilateral $ABCD$ with vertices $A(-1,2)$, $B(1,3)$, $C(-5,10)$ and $D(-3,11)$ is a rectangle. In order to do this, show that two pairs of opposite sides are parallel. Then show that one angle of the quadrilateral is a right angle.

6.9 THE MEASURE OF A SEGMENT

We may think of the numbers on the x-axis and the y-axis as denoting linear units of measure.

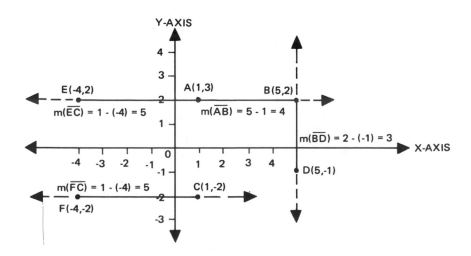

Figure 6.25

By counting, we can determine that in Figure 6.25 the measure of a line segment containing point A and B is 4; containing B and D is 3; and containing C and F is 5. It is apparent that the two points A and B have the same ordinate, 2. We can then find the measure of a line segment containing points A and B by finding the difference of their abscissas, or $5-2 = 3$. Similarly, we could find the measure of a line segment containing points A and E by finding the difference of their abscissas, or $1-(-4) = 5$. The reader should verify, by counting, that these measures are correct.

If we consider the points B and D of Figure 6.25, we see that the two points have the same abscissa, 5. We may find the measure of a line segment containing these two points by subtracting their ordinates, or $2-(-1) = 3$. By counting, the reader may verify this measure.

If we wish to find the measure of a line segment containing two points with unequal abscissas and unequal ordinates, we resort to the Pythagorean Theorem by first "constructing" a right triangle. Recall that the Pythagorean Theorem states that "the square on the hypotenuse of a right triangle is equal to the sum of the squares on the two sides of the triangle."

Referring to Figure 6.26, if we construct a line segment parallel to the y-axis that includes point B and a line segment parallel to the x-axis that contains point A, we find these two line segments intersect at one point, Q. Since $\overline{BQ} \perp \overline{AQ}$, the three points A, B and Q determine the right triangle $\triangle ABQ$.

Using the measures of the sides of this triangle, we may determine the measure of $\overline{AB}$, using the Pythagorean Theorem:

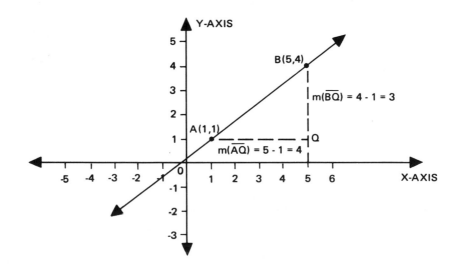

Figure 6.26

$$m(\overline{AB}) = \sqrt{3^2 + 4^2}$$

$$= \sqrt{9 + 16}$$

$$= \sqrt{25}$$

$$= 5$$

Hence, we have a method which we can use to determine the measure of any segment that "slopes uphill" or "slopes downhill." If we consider the segment containing $P_1(x_1,y_1)$ and $P_2(x_2,y_2)$ in Figure 6.27, we can easily determine the formula for finding $m(\overline{P_1 P_2})$. Using the Pythagorean Theorem, the measure of the segment with end-points $P_1(x_1,y_1)$ and $P_2(x_2,y_2)$ can be stated as:

$$m(\overline{P_1 P_2}) = \sqrt{(x_2 - x_1)^2 + (y_2 - y_1)^2}$$

In most textbooks, this is referred to as the "distance formula," and $m(\overline{P_1 P_2})$ is stated as "the distance between two points."

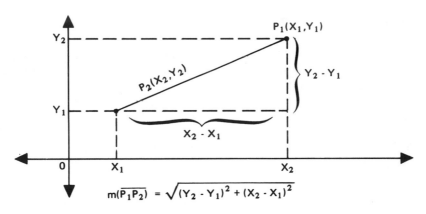

Figure 6.27

Problem and Activity Set 6.8

1. Without using the distance formula, state the distance between each pair of points.
   a. $A(1,4)$ and $B(6,4)$ b. $A(-3,2)$ and $B(7,2)$
   c. $A(2,0)$ and $B(2,6)$ d. $A(0,1)$ and $B(0,7)$
2. Write a simple formula for the distance between $A(x_1,k)$ and $B(x_2,k)$. (Hint: The points are members of a horizontal line).
3. Write a simple formula for the distance between $A(k_1,y_1)$ and $B(k,y_2)$.

4. Use the distance formula to find the distance between
   a. $(0,0)$ and $(1,5)$ b. $(0,0)$ and $(-1,-4)$
   c. $(1,2)$ and $(6,14)$ d. $(10,1)$ and $(3,2)$
   e. $(8,11)$ and $(15,35)$ f. $(-6,3)$ and $(4,-2)$
5. Use the distance formula to simplify the statement: The square of the distance between $(0,0)$ and (x_1,y_1) is 49.
6. Show that the triangle ABC with vertices $A(0,0)$, $B(3,4)$, and $C(-1,1)$ is isosceles by computing the lengths of the sides.
7. Vertices W, X, and Z of rectangle $WXYZ$ have coordinates $W(0,0)$, $X(2,0)$, and $Z(0,3)$, respectively.
   a. What are the coordinates of Y?
   b. Using the distance formula, show that $m(\overline{WY}) = m(\overline{XZ})$.
8. Find the lengths of the sides of quadrilateral $ABCD$ with vertices $A(6,4)$, $B(3,-2)$, $C(-1,2)$ and $D(2,8)$.
9. Find the length of each diagonal of quadrilateral $ABCD$ of Problem 8.
10. Show that quadrilateral $ABCD$ with vertices $A(2,1)$, $B(2,-1)$, $C(4,-1)$ and $D(4,1)$ is a square. Show that the diagonals of $ABCD$ are perpendicular.

6.10 EQUATIONS OF CIRCLES

Recall that a circle was defined as a set of points whose measure from a given point A is r linear units. In Figure 6.28 let A be the origin and $r = 2$. The circle is illustrated in the coordinate plane.

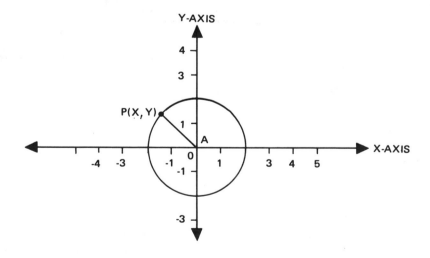

Figure 6.28

If we use the "distance formula" and denote the points of the circle in Figure 6.28 as $P(x,y)$, then

$$2 = \sqrt{(x-0)^2 + (y-0)^2}$$

or

$$2 = \sqrt{x^2 + y^2}$$

If we square both members of this equation, we have

$$4 = x^2 + y^2$$

The truth set of this equation are points of the circle.

If we let r be the measure of the radius of the circle, the equation $x^2 + y^2 = r^2$ is the equation of a circle whose center is the origin and the radius r linear units.

Problem and Activity Set 6.9

1. The circle shown has a radius of five units. Find the value of:
   a. $x_1^2 + y_1^2$
   b. $x_3^2 + y_3^2$
   c. $0^2 + y_4^2$
   d. $x_5^2 + y_5^2$
   e. $x_6^2 + 0^2$

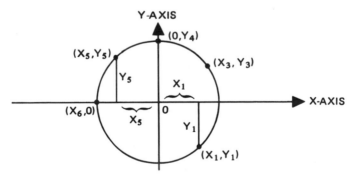

2. Determine the radius of each of the following circles. Draw the circles on graph paper.
   a. $x^2 + y^2 = 4^2$
   b. $x^2 + y^2 = 25$
   c. $x^2 + y^2 = 4$
   d. $x^2 + y^2 = 9$
3. Find the equations of the following circles and draw their graphs.
   a. Center $(0,0)$, radius 5
   b. Center $(0,0)$, radius 6
   c. Center $(0,0)$, radius 3

4. Each circle below has center (0,0) and contains the given point. Find the equation of the circle. Draw a graph of the circle.
   a. Circle P contains point (4,3)
   b. Circle Q contains point (12,5)
   c. Circle R contains point (−6,8)

SELECTED ANSWERS
FOR PROBLEM AND ACTIVITY SETS

CHAPTER 1

Problem Set 1.1

1. b. $\{$March, May$\}$.
4. c. True.   e. True.
   g. False.
8.

10. b. *B, H.*   c. *C, E, G.*
11. Ex: $\{\ \ 5, 10, 15, 20, 25, \ldots\}$
    $$\updownarrow \ \updownarrow \ \updownarrow \ \updownarrow \ \updownarrow$$
    $\{10, 20, 30, 40, 50, \ldots\}.$
13. b. Ex: $\{$June, July$\}$.
16. $\{1, 2, 3\}, \{1, 2\}, \{1, 3\}, \{2, 3\}, \{1\}, \{2\}, \{3\}, \{\ \ \}.$ 7.
18. a. $E = A.$ b. No other sets are equal to *C.*
    c. No.   d. $D = F.$
21. a. True.   b. True.
    c. False.

23. $\{2, 3, 4, 6, 8, 9, 10, 12\}$.
27. D.
29. $\{3\}; \{5\}; \{ \}; \{3, 5, 7, 13, 15\}; \{3, 5, 7, 9, 11\}; \{3, 5, 9, 11, 13, 15\}$.
34. A.
36. $\{1, 2, 6, 7\}$.

CHAPTER 2

Problem Set 2.1

1. D.
4. C.
7. More points than we can count.
9. More paths than we can count.
11. $3; \overleftrightarrow{AB}, \overleftrightarrow{BC}, \overleftrightarrow{AC}$.

18. $\overline{AB}; \overline{AC}; \overline{AD}; \overline{BC}; \overline{BD}; \overline{CD};$ 6.
23. b. True. ●———● ; $\overline{AB} = \overline{BA}$.
            A   B
    d. False.
    g. True. ◄●————●————●► $\overline{AB}, \overline{BC}, \overline{AC}$.
            A   B   C

Problem Set 2.2

2. a. True.   b. False.
   c. True.   d. True.
   e. False.   f. True.
4. Point A.
8. $\{ \}; \{ \}; D$.

Problem Set 2.3

1. More points than can be counted.
4. No.
8. $\{ \}$.
10. c. Ex: $\{$Plane $ABH \cap$ Plane $DBI \cap$ Plane $CDB = B$.
    e. Ex: $\overleftrightarrow{EJ} \cap \overrightarrow{BH} = \{ \}$.
13. a. Exactly 1 point or is $\{ \}$.
    b. No. $AB \cap \ell$ may also be $\{ \}$.
    c. Yes.

Problem Set 2.4

1.

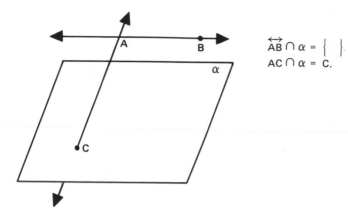

$\overleftrightarrow{AB} \cap \alpha = \{\ \ \}.$
$AC \cap \alpha = C.$

5. c. Ex: $\overleftrightarrow{DC}$ and $\overleftrightarrow{FG}$.

  Ex: Plane $HDC \cap$ Plane $GCB \cap$ Plane $DCB = C$.

Problem Set 2.5

2. Yes; The axle; The clock hands.
3. $A; \overrightarrow{AB}, \overrightarrow{AC}. \ E; \overrightarrow{ED}, \overrightarrow{EF}. \ H; \overrightarrow{HG}, \overrightarrow{HI}. \ K; \overrightarrow{KJ}, \overrightarrow{KL}.$
5. a. One.   b. One.
  c. One. $\angle BAC$ or $\angle CAB$.
9. b. $\{\ \ \}.$ f. Yes.
  h. No.
11. c. Ex:

  $\angle A \cap \angle B =$ Exactly 2 points.

Problem Set 2.6

1. $\overline{CD} \cong \overline{AB}.$
2. $\angle OMN \cong \angle ABC.$
4. $A \cong H \cong D; \ B \cong T; \ C \cong J \cong P; \ E \cong R; \ F \cong Q; \ I \cong K; \ G \cong S; \ L \cong N.$

Problem Set 2.7

1. a. $\angle CAD$ and $\angle BAE.$ b. $\angle CAB$ and $\angle EAD.$
  c. $\angle CAB.$ d. $\angle DAC$ and $\angle EAB$ are vertical $\angle$'s.

3. b. 8 pairs.
   d. No. $\ell_1 \cap \ell_2 = \{\quad\}$.
   g. 4 pairs. $\angle CAH$ and $\angle GHF$; $\angle DAB$ and $\angle AHE$; $\angle BAH$ and $\angle EHF$;
      $\angle DAC$ and $\angle AHG$.
5. a. ℓ_3 is the transversal.   b. Ex: $\angle B$ and $\angle F$.
   c. Ex: $\angle C$ and $\angle G$.   d. $\angle C$ and $\angle E$; $\angle D$ and $\angle F$.
   e. $\angle A \simeq \angle C \simeq \angle E \simeq \angle G$.   f. $\angle B \simeq \angle D \simeq \angle F \simeq \angle H$.

Problem Set 2.8

2. $\angle HAD$, $\angle HAC$, $\angle HAB$.
   a. $\angle BAF$ and $\angle JAF$.
3. b. E.   d. F.
   f. B.
6. b. False.   g. False.
   j. True.   l. True.
   q. False.

CHAPTER 3

Problem Set 3.1

1. $m(\overline{CD}) = 2$ glubs.
4. No two.
6. a. $32/8$.
8. AC, 7 inches; CB, 7 inches; BD, 4 inches; DA, 4 inches.
10. d. False.

Problem Set 3.2

2. $m(\angle A) = 10$.
3. j. $\angle AHC$, $\angle CHI$; $\angle AHD$, $\angle DHI$; $\angle IHE$, $\angle EHA$.
5. $45, \angle BPD, \angle DPC$.
9. a. Yes, corresponding angles; Yes, congruent angles.
13. b. 30.
19. $A = 60$, $B = 120$, $C = 60$.

CHAPTER 4

Problem Set 4.1

1. Ex: a hula-hoop; the border of a table cloth.
2. $C, H, M,$ and P.
6. No. Yes. Two simple closed curves.

9. Ex:

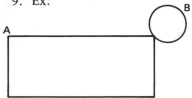

19. 2; $\overline{AB}$, $\overline{AC}$ and $\overline{CB}$; $\overline{CB}$, $\overline{CD}$, and $\overline{DB}$. 5.

Problem Set 4.2

4. b. True.   f. True.
7. Interior $X \cap$ Interior Y.

Problem Set 4.3

2. a. ... triangle ABC.   b. acute.
   c. acute.   d. acute.
14. ⊀2 ≃ ⊀3; ⊀1 ≃ ⊀4; ⊀5 ≃ ⊀8; ⊀6 ≃ ⊀9; ⊀7 ≃ ⊀10; ⊀11 ≃ ⊀13; ⊀12 ≃ ⊀14.

Problem Set 4.4

2. ⊀BAD and ⊀ADC; ⊀ADC and ⊀DCB; ⊀DCB and ⊀CBA; ⊀CBA and ⊀BAD.
   ⊀A and ⊀C; ⊀D and ⊀B.
5. 0, 2, 5, 9.

Problem Set 4.5

2. a. 3.   b. 1.
   c. 1.   d. 1.
   e. 2.
3. a. $\overline{JK}$ and $\overline{AC}$.   d. $\overline{JE}$ and $\overline{HF}$.
   f. $AJED$; $AHFD$; $JHFE$; $ICFH$.

Problem Set 4.6

2. 8, 2.
10. 8 sq. in.
13. a. 100.

Problem Set 4.7

1. a. 56.
2. c. 540.

Problem Set 4.8

1. b. 480, 416.

5. 500.
9. b. 16,000.

CHAPTER 5

Problem Set 5.1

1. a. Ex: $\overline{OE}$; $\overline{OC}$.   b. Interior.
    c. O, A, B, C, D, E, F.   d. Yes.
4. d.

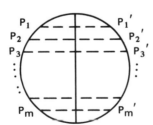

8. g. That portion of the circle between G and H that are also interior points of circle B.

Problem Set 5.2

1. a. 3.   b. 4.
    c. 5.
2. b. $\overline{BD}$.   c. $\overline{AP}, \overline{PC}, \overline{PB}, \overline{PD}$.
8. b. False.   d. True.

Problem Set 5.3

1. a. $\overset{\frown}{EDC}$.   f. $\overset{\frown}{DEF}$.
4. a. 180.   b. approximately 336.
8.

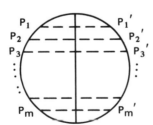

Problem Set 5.4

1. a. $m(P) = 37.68$; $m(A) = 112.04$.
    c. $m(P) = 94.20$; $m(A) = 706.50$.
5. Skillet.
7. $15/2\pi$.

Problem Set 5.5

1. a. Yes.
   c. One.
   e. No.
   g. No. It is possible that the intersection contain no points.

   b. No.
   d. So many we cannot count them.
   f. Yes.

6. a. 100°E, 40°N.
   c. 150°W, 50°S.

Problem Set 5.6

1. a. $A \approx 5,040$ sq. in.; $V \approx 33,400$ cu. in.
   b. $A \approx 113$ sq. ft.; $V \approx 113$ cu. ft.

CHAPTER 6

Problem Set 6.1

1. a. $(\overset{\rightarrow}{2}, 2\uparrow)$.
   f. $(\overset{\leftarrow}{2}, 3\uparrow)$.

   c. $(0, 2\downarrow)$.

Problem Set 6.2

3. $A(3, 1)$; $B(-2, -5)$; $C(4, -2)$.

Problem Set 6.3

1. a. approximately 17°.
   c. 90°.
   e. approximately 33°.
2. a. ? = 2.
6. a. ? = 1.
9. b. −3/2.

   b. approximately 153°.
   d. 0°.
   f. approximately 135°.

Problem Set 6.4

1. a.

x	y
0	5
1	$9/2$
4	3
10	0
−1	$11/2$
14	−2

b.

x	y
10	0
5	10
11	−2
15	−10
−7	4
8	4

Problem Set 6.5

1. a. $4x - y = 3$.   c. $5x - y = -5$.
4. a. $y + 2x = 4$.   c. $y + 3x = 3$.

Problem Set 6.6

1. a. 4.   c. -6.
   e. 0.   g. -4.
   j. y-axis.
2. b. $x + y = 1$.   e. $x = 0$.
   g. $x = 2$.

Problem Set 6.7

1. a. $(2, 4)$.   b. No common members.
   c. No common members.
4. Slope of $\overline{AC} = 1$; Slope of $\overline{CB} = -1$.

Problem Set 6.8

2. $m(\overline{AB}) = |x_2 - x_1|$.
4. b. $\sqrt{17}$.   e. 25.
6. $m(\overline{AB}) = m(\overline{BC}) = 5$.
9. $m(\overline{DB}) = \sqrt{101}$; $m(\overline{AC}) = \sqrt{45}$.

Problem Set 6.9

1. b. 25.   e. 25.
3. b. $x^2 + y^2 = 36$.
4. b. $x^2 + y^2 = 169$.